Lecture Notes in Mathematics

Edited by A. Dold and B. Eckmann

1264

AF001646

Wu Wen-tsün

Rational Homotopy Type

A Constructive Study via the Theory of the I*-measure

Springer-Verlag

Berlin Heidelberg New York London Paris Tokyo

Author

Wu Wen-tsün
Institute of Systems Science, Academia Sinica
Beijing, 100080, P.R. China

Mathematics Subject Classification (1980): 57XX, 55XX, 58XX

ISBN 3-540-13611-8 Springer-Verlag Berlin Heidelberg New York
ISBN 0-387-13611-8 Springer-Verlag New York Berlin Heidelberg

This work is subject to copyright. All rights are reserved, whether the whole or part of the material is concerned, specifically the rights of translation, reprinting, re-use of illustrations, recitation, broadcasting, reproduction on microfilms or in other ways, and storage in data banks. Duplication of this publication or parts thereof is only permitted under the provisions of the German Copyright Law of September 9, 1965, in its version of June 24, 1985, and a copyright fee must always be paid. Violations fall under the prosecution act of the German Copyright Law.

© Springer-Verlag Berlin Heidelberg 1987
Printed in Germany

Printing and binding: Druckhaus Beltz, Hemsbach/Bergstr.
2146/3140-543210

P R E F A C E

D. Sullivan has discovered the following remarkable fact: The differential graded algebra (abbreviated DGA) of differential forms on a differential manifold contains information not only on the infinite part of the cohomology ring of the manifold, but also on the infinite part of the homotopy ring under the Whitehead product. It was also D. Sullivan who extended the notion of DGA of differential forms to an arbitrary simplicial complex. Moreover, he introduced the notion of minimal model for such a DGA, and showed that the same relations to homology and homotopy hold as in the case of manifolds; these relations can easily be deduced from such a minimal model. Though the main ideas of the theory may be traced back, as shown by Sullivan himself, to many predecessors from E. Cartan onwards, it is Sullivan who is the uncontestable founder of this whole theory which immensely influences further developments of algebraic topology. The present book aims at presenting the theory in an elementary form under the name of I^*-measure which is a synonym of "rational homotopy type" or "minimal model" in current use. We adopt this name to give evidence of its measure-theoretical character, with particular emphasis on its calculability. In fact Sullivan has expressed the opinion that one of the advantages of this theory lies in offering "a new concreteness for calculation". We introduce here the notion of calculability of a certain measure with respect to a certain geometrical construction in a technical sense. It is one of the aims of this book to show that the I^*-measure is calculable with respect to most of the important geometrical constructions usually encountered in algebraic topology, in contrast to the fact that even the simplest classical cohomology H^*-measure is not calculable, even with respect to the simplest geometrical constructions such as the addition (i.e., union) and multiplication (i.e., product-formation of spaces). Quite often explicit formulae exhibiting such calculability will be given, which put in evidence the concreteness of calculation for the measure in question.

The book is divided into seven chapters. Chapter I, which is introductory, introduces the basic notions of measure and its calculability. Chapter II, of a purely algebraic character, is concerned with the notion of differential graded algebras over a field $\underline{k}$ of characteristic 0, to be abbreviated as DGA/$\underline{k}$ or simply DGA, with the homology of a DGA, as well as the homotopy of DGA-morphisms; and, what is of utmost importance, with the notion of minimal model of a DGA $\mathcal{A}$, denoted as min $\mathcal{A}$, and various properties connected therewith. Chapter III introduces the notion of the deRham-Sullivan measure $A^*(K)$, the DGA of differential forms on a simplicial complex K, and the fundamental notion of I^*-measure $I^*(K)$ of K, defined simply as the minimal model of $A^*(K)$: $I^*(K) = \min A^*(K)$. Moreover, in Chapter III the homotopy invariance of $I^*(K)$ is proved as well as the generalized deRham theorem due to Sullivan, viz.

$$H(I^*(K)) \approx H(A^*(K)) \approx H^*_{\underline{k}}(K) ,$$

which shows that $I^*(K)$ contains complete information on the cohomology of K with coefficients in a field $\underline{k}$ of characteristic 0 . Chapter IV shows that the I^*-measure $I^*(K)$ also contains complete information about the homotopy ring of K und the Whitehead product after tensoring with the basic field $\underline{Q}$ of rationals. Thus Chapters III and IV together give the fundamentals of the theory of I^*-measure in showing its connection with the classical homology and homotopy measures of a space. Chapter V gives a concrete example of actually calculating the I^*-measure in the important case of a homogeneous space. It begins with the work of E. Cartan and ends with an extension of a theorem of H. Cartan on real cohomology to the case of an I^*-measure. The known published proofs of this theorem are quite complicated; the proof given here follows closely the paper of Rashevskii which seems to be somewhat simpler. Chapter VI deals with the calculability of I^*-measure with respect to addition (or union) of two complexes, which is not so evident, in contrast to the multiplication (or product-formation) of two complexes which is entirely trivial. In this Chapter we give some explicit formulae for the determination of the I^*-measure of the resulting complex in terms of those of the component complexes, as well as the interrelated morphisms. As a consequence we show how to calculate, at least theoretically, the I^*-measure of a finite complex in an algorithmic manner, once the combinatorial structure of the complex is known. In this way we also establish an axiomatic system for the I^*-measure of finite complexes in a manner quite different from that of the well-known Eilenberg-Steenrod axiomatic system for H^*-measures. As applications of the method developed in this Chapter we prove a theorem about the explicit determination of the I^*-measure of the fiber space in terms of those of the base and the fiber, for a fibration $F \subset E \to B$; namely,

$$I^*(E) \approx \min(I^*(B) \otimes_\tau I^*(F)) .$$

In taking homology of both sides, we then get

$$H^*_{\underline{k}}(E) \approx H(I^*(B) \otimes_\tau I^*(F)) ,$$

in which I^* cannot be replaced by the cohomology H^* unless very special conditions are imposed on the base or the fiber. This furnishes a concrete example of showing clearly the superiority of the I^*-measure over the classical H^*-measure in the case where the coefficient field $\underline{k}$ is of characteristic 0 . In the formulae above $\otimes_\tau$ means that the differential in the tensor product is twisted. The way of arriving at those formulae also shows clearly how such twisted differentials occur. The twisted differential can be explicitly determined in many important cases described in this and the next chapter. In the final Chapter VII various spectral sequences connected with fibrations are studied; they lead to an explicit determination of the I^*-measure of a fibre-square-constructed space. As an application we show that for a fibration with a homogeneous-space fiber with the usual structure group, the I^*-measure of the fiber space will be completely determined by those of the base and the fiber, and by certain characteristic elements which, when passing to homology, are just characteristic classes of the fibration in the usual sense.

Mathematics should be constructive. Therefore, in the treatment chosen in this book we stick as much as possible to a constructive point of view. Thus we impose in Chapter I three requirements for a measure to be efficient which include, besides invariance and calculability some kind of finiteness. The spaces considered will be restricted accordingly, to those having the homotopy type of a countable connected simplicial complex in the weak topology, although all the concepts and results are valid for much more general spaces. In Chapter III the proof of the deRham-Sullivan theorem is based on the one given originally by Weil with appropriate modifications to remove all traces of non-constructive character, so that the final isomorphism in the theorem is explicit and constructive. For the same reason in Chapter IV we have used the Cartan-Serre tower of more or less constructive nature, instead of the Postnikov tower of mere existential character. Such constructiveness is quite evident in our treatment in Chapters V and VI, and even in part of Chapter VII.

A brief outline of the present book has appeared in The Chern Symposium 1979. Essential part of it constituted the contents of a course on algebraic topology given at Berkeley in the spring semester of 1981. The author would like to express his gratitudes to many mathematicians of the United States, France and West-Germany for their invitation to visit their respective universities or institutes. They are too numerous to be all mentioned here but among them the author would like to mention in particular Prof. S.S. Chern of UC at Berkeley, Prof. A. Weil and Prof. J. Milnor of IAS, Prof. Kuiper of IHES, and Prof. Hirzebruch of MPI. The author has benefited greatly and in many respects from these relatively long periods of stay. For example, it was during his stay at IHES that he had the possibility to make personal contact with Prof. D. Sullivan and that he had paid a visit to the University of Lille, a center of development of Sullivan's theory; there the author was given a collection of Lille Publications, which was very helpful to his later research. It was also during his stay at MPI that he made acquaintance with Prof. H.J. Baues who was kind enough to give him access to his personal collection of published or unpublished literature on rational homotopy theory. Of particular importance to the author was the visit of Profs. W. Browder, J. Spencer and F. Peterson to China in the spring of 1973. They brought us a copy of the Tokyo Symposium on Manifolds, and after their return to the USA Prof. F. Peterson has sent us a xerox copy of the Italian lecture notes on Sullivan's theory by Friedlander, Griffiths and Morgan, at that time yet hardly known even to the western mathematical world. Thanks to these precious presents the author became aware of and attracted by this beautiful theory and started his own research in the field.

Most of the results in this book have appeared in some form in the works of D. Sullivan. However, they have been clarified, generalized, discovered and rediscovered, formulated and reformulated by many others in various different presentations. It is indeed difficult to ascertain who should be considered the real contributor of this or that result, so that we have left aside all such questions of priority. Furthermore, the theory is now well-developed and has many ramifications, extensions, and applications. The author is quite unable to touch on these. In fact, he had to limit himself to a presentation of the most fundamental part of the theory centered around his own studies from his own point of view. In particular, the author very much regrets not even being able to give a sketch of the important contributions of Prof. K.T. Chen who may be considered as a co-founder of the theory, in quite a different setting. The bibliography will also be restricted to those items which are strongly related to the present text. However, besides the Italian lecture notes cited above, the author would like to mention in particular the following works which have been of great help, both for his research and for his writing:

1. S. Halperin: Lectures on minimal models, Lille Publications (1978 and 1981).
2. D. Lehmann: Théorie homotopique des formes différentielles d'après Sullivan, Astérisque (1977).
3. P. Déligne, P. Griffiths, J. Morgan and D. Sullivan: The real homotopy theory of Kähler manifolds, Invent. math. 29(1975), 245-254.
4. D. Sullivan: Infinitesimal computations in topology, Publ. Math. 47(1977), 269-331.

Finally, the author would like to express his gratitude to Mrs. Hermona Rosinger for her excellent typing. The author is also very grateful to the staff of Springer-Verlag in admitting this work to their publication list. Last but not least, the author expresses his gratitude to his wife, Chen Pi-ho; without her assistance in so many respects the writing of this book would have been impossible.

C O N T E N T S

C h a p t e r   I

FUNDAMENTAL CONCEPTS. MEASURE AND CALCULABILITY

I.1 THE NOTION OF MEASURE

Definition 1: A geometrical category GEOM is just a collection of geometrical objects of a certain kind. A geometrical construction CONS, which permits to construct in a definite manner a new geometrical object O from a finite set of geometrical objects $O_1, \ldots, O_n$, is said to be pertaining to the geometrical category GEOM, if the new object O belongs to GEOM whenever $O_1, \ldots, O_n$ belong to GEOM.

In other words, CONS is pertaining to GEOM if GEOM is closed with respect to CONS.

Definition 2: An algebraic category ALG is just a collection of algebraic objects of a certain kind. An algebraic construction CONS, which permits to construct in a definite manner a new algebraic object O from a finite set of algebraic objects $O_1, \ldots, O_n$, is said to be pertaining to the algebraic category ALG if the new object O belongs to ALG whenever $O_1, \ldots, O_n$ all belong to ALG. In other words, CONS is pertaining to ALG if ALG is closed with respect to CONS.

Definition 3: Given a geometrical category GEOM and an algebraic category ALG a correspondence Meas which associates to any object O in GEOM a definite object in ALG will be said to be an ALG-measure on GEOM or simply a measure on GEOM. In notation:

Meas: GEOM $\rightsquigarrow$ ALG.

For the usual algebraic categories we may mention the following trivial ones with the ordinary addition and multiplication as pertaining algebraic constructions:

$\underline{Z}$ = Collection of all integers,

$\underline{R}$ = Collection of all real numbers,

$\underline{C}$ = Collection of all complex numbers,

$\underline{Q}$ = Collection of all rational numbers

$A[t_1, \ldots, t_n]$ = Collection of all polynomials in indeterminates $t_1, \ldots, t_n$ with coefficients in A, a given commutative ring, etc.

In the examples below the algebraic category will not be specified.

Ex. 1:

GEOM = Collection of all polygons (connected or disconnected) in a euclidean plane.

CONS = Rigid motion, union, or intersection, etc.

Meas = Area of a polygon.

Similarly for collection of polyhedra in a 3-space as GEOM and volume as Meas.

Ex. 2:

GEOM = Collection of all bounded point sets in a plane.

CONS = Rigid motion, union, intersection, or formation of derived sets, etc.

Meas = Content or Lebesgue measure of a point set.

Ex. 3:

GEOM = Collection of all polygonal knots in a euclidean 3-space.

CONS = Certain allowable deformations, connected sum or amalgamation of two knots into one, etc.

Meas = Genus, signature, Alexander polynomial, or π_1 of the complement of a knot, etc.

Ex. 4:

GEOM = Collection of all finite linear graphs.

CONS = Formation of dual graph in the case of planar graphs, amalgamation of two graphs into one, etc.

Meas = Incidence matrix, chromatic polynomial, thickness, etc.

Ex. 5:

GEOM = Collections of various kinds of topological spaces, in particular complexes, manifolds, etc.

CONS = Formation of union and product of spaces, formation of universal covering space, shrinking of a subspace to a point, surgery in case of manifold, etc.

Meas = Dimension, Euler characteristic, betti numbers, etc. whenever they are well-defined.

Besides these numerical measures we encounter in algebraic topology various kinds of measures in the form of much more complicated algebraic structures like groups, rings, differential graded algebras, etc. Thus, we have:

Singular chain or cochain groups;

Homology or cohomology groups or rings over certain coefficient groups or rings;

Homotopy-group in various dimensions;

K-ring and various generalized cohomology rings, etc.

To put in evidence their measure character we shall adopt the following notations somewhat deviated from the usual ones (X = space in the given geometrical category, G = commutative group, A = commutative ring):

$H_n^G(X)$ instead of $H_n(X,G)$;

$H_G^n(X)$ instead of $H^n(X,G)$;

$H_\oplus^G(X)$ instead of $H_*(X,G)$ as a group;

$H_G^\oplus(X)$ instead of $H^*(X,G)$ as a group;

$H_A^*(X)$ instead of $H^*(X,A)$ as a ring;

etc.

I.2 EXAMPLES IN MEASURES WITH APPLICATIONS

Ex. A. The Brouwer fixed-point theorem

Let D^n be the closed unit disc in the euclidean n-space and $f : D^n \to D^n$ any map of D^n into itself. The Brouwer fixed-point theorem asserts the existence of a fixed point $x \in D^n$, such that $f(x) = x$.

We shall prove it by means of certain appropriate measures on some appropriate geometrical categories as follows.

Suppose the contrary, that no such fixed point exists. Let the boundary sphere of D^n be $S = S^{n-1}$. For any point $x \in S$ and any real number $t \in [0,1]$ the points tx and $f(tx)$ are then different, and their joining line in the direction from $f(tx)$ to tx will meet S in a unique point, say $g_t(x)$. For $0 \leq t \leq 1$ the maps

$$g_t : S \to S$$

give then a homotopy between the constant map g_0 and the identity map g_1.

Consider now a geometrical category GEOM of spaces containing $S = S^{n-1}$ as a particular one and a group-measure M on GEOM verifying the following properties, which may also be considered as axioms about the measure in question:

A1. Every map

$$h : X \to Y$$

of spaces X, Y in GEOM will induce a morphism

$$h_M : M(X) \to M(Y) .$$

A2. As in A1 for maps homotopic to each other

$$h \simeq h' : X \to Y$$

the morphisms induced will be identical:

$$h_M = h'_M : M(X) \to M(Y) .$$

A3. For the particular space S the measure $M(S) \neq 0$ (i.e. $M(S)$ is non-trivial in the corresponding algebraïc category).

A4. For a particular constant map $h : S \to S$ the induced morphism is

$$h_M = 0 \quad (\text{i.e. } h_M(M(S)) \text{ is trivial or } = 0).$$

A5. For the particular identity map $h : S \to S$ the induced morphism is

$$h_M = \text{Identity} .$$

It is clear, that there will result some contradiction from the properties A1 – A5, if we assume, that no fixed point of f exists. Consequently the Brouwer theorem is true, and the above reasoning furnishes a proof in assuming the existence of a measure M verifying the properties A1 – A5. For such measures we may take e.g. H^G_{n-1} or π_{n-1}. Even π_n or π_{n+1} may be used as such measures in case $n \geq 3$, since $\pi_n(S^{n-1}) \approx \underline{Z}$ or $\underline{Z}_2$ and $\pi_{n+1}(S^{n-1}) \approx \underline{Z}_2$ verify the required properties.

The measures above are <u>covariant</u> in character, in that the induced morphism

$$h_M : M(X) \to M(Y) \quad \text{of a map} \quad h : X \to Y$$

is in the same direction as h from X to Y. We may also consider measures <u>contravariant</u> in character, with morphism induced by a map $h : X \to Y$ now in reverse direction as h, being from Y to X. It will then be denoted as

$$h^M : M(Y) \to M(X) .$$

The same reasoning may be applied to prove the Brouwer fixed point theorem, in assuming the existence of such a contravariant measure, verifying properties analogous to A1 – A5 with, however, all directions of morphisms to be reversed. The usual cohomology-group H^{n-1}_G is then an example of such a measure.

We may even take the integer measure M with $M(S) = \underline{Z}$ and

$$h_M \quad (\text{or } h^M) : M(S) \to M(S)$$

for $h : S \to S$ given by

$$h_M(m) \quad (\text{or } h^M(m)) = d(h) \cdot m , \qquad m \in \underline{Z} ,$$

where $d(h)$ is some integer, depending on the map h, to be called the <u>degree</u> of h. With the geometrical category GEOM consisting of a single space S and properties A1 – A5 we get again a proof of the theorem, which is in essence the original one of Brouwer.

Ex. B. Fundamental Theorem of Algebra

Let C be the complex plane and

$$f : C \to C$$

the map given by $(z \in C)$

$$f(z) = z^n + a_1 z^{n-1} + \dots + a_n$$

with $a_i \in C$. We shall prove, that $f(z) = 0$ for some $z \in C$ by means of certain appropriate measures on some appropriate geometrical category.

For this purpose let us set $a = \mathrm{Max}(|a_i|, 1)$ and take some $R > na$. Set

$$S = S(R) = \{z \in C / |z| = R\},$$
$$D = D(R) = \{z \in C / |z| \leq R\}.$$

There is then a radial projection

$$\pi = \pi(R) : C - (0) \to S.$$

Suppose that the theorem is not true, so that $f(z) \neq 0$ for all $z \in C$. Then

$$\pi f : D \to S$$

is well defined. For $0 \leq t \leq 1$ define a map

$$f_t : C \to C$$

by

$$f_t(z) = z^n + t(a_1 z^{n-1} + \dots + a_n)$$

with

$$f_1 = f.$$

For $0 \leq t \leq 1$ and $z \in S$ we have

$$|f_t(z)| \geq |z^n| - t(|a_1|\,|z|^{n-1} + \ldots + |a_n|) \geq R^n - na\,R^{n-1} > 0 \ .$$

Hence

$$\pi f_t : S \to S$$

is well-defined for any $0 \leq t \leq 1$ and defines a homotopy

$$\pi f \simeq g : S \to S$$

where g is given by

$$g(z) = z^n / R^{n-1} .$$

On the other hand, the map

$$\pi f'_t : S \to S$$

given by

$$\pi f'_t(z) = \pi f(tz)$$

for $z \in S$ is also well-defined for any $0 \leq t \leq 1$ and defines a homotopy

$$\pi f \simeq g' : S \to S$$

where

$$g'(z) = \pi f(0)$$
$$= \text{a constant} \ .$$

Suppose now, there is some geometrical category GEOM of spaces containing the circles S as particular ones, and a certain non-trivial group-measure M on GEOM, verifying properties B1 – B5 below:

B1 – B4. Same as A1 – A4 in Ex. A.

B5. For the particular map

$$g : S \to S$$

given by

$$g(z) = z^n / R^{n-1}$$

the induced morphism g_M is given by

$$g_M(m) = nm \; , \qquad m \in M(S)$$

with $nm \neq 0$ for any $m \neq 0$ in $M(S)$.

It is clear, that a contradiction will result from the properties B1 – B5. Consequently the fundamental theorem of algebra is proved in assuming the existence of a group-measure M, verifying the above properties. For such measure we may take e.g. the homology group H_1^G for any abelian group G or the fundamental group π_1 . We may also use the cohomology group H^1_G as some contravariant measure, verifying properties analogous to B1 – B5, with directions of morphisms reversed. We may also use some degree measure, as indicated at the end of Ex. A.

Ex. C. <u>Theorem: The n-sphere S^n cannot be imbedded in the euclidean n-space $\underline{R}^n$</u> .

To prove this, suppose that the theorem is not true, and

$$f : S^n \subset \underline{R}^n$$

is an imbedding. Then $f(x) \neq f(-x)$ for all $x \in S^n$ with $-x$ the antipodal point of x. Let S^{n-1} be the unit (n-1)-sphere in $\underline{R}^n$. It is easy to deform f to a map

$$\tilde{g} : S^n \to S^{n-1} \; ,$$

which maps antipodal points of S^n to antipodal points of S^{n-1}. In this way we get a commutative diagram of maps

$$\begin{array}{ccc} S^n & \xrightarrow{\tilde{g}} & S^{n-1} \\ {\scriptstyle p_n}\downarrow & & \downarrow{\scriptstyle p_{n-1}} \\ P^n & \xrightarrow{g} & P^{n-1} \end{array}$$

in which P^n, P^{n-1} denote projective spaces and p_n, p_{n-1} the corresponding covering maps.

Now, suppose that there exists a certain geometrical category GEOM of spaces, containing spheres and projective spaces as particular ones, as well as a contravariant measure M on GEOM in the algebraic category of algebras over $\underline{Z}_2$, verifying the following properties:

C1. Each map of spaces in GEOM

$$h : X \to Y$$

will induce a morphism of measures

$$h^M : M(Y) \to M(X) .$$

C2. If $\tilde{X}$ is a two-sheeted covering space of X and

$$p : \tilde{X} \to X$$

is the corresponding covering map, then there is a certain <u>characteristic element</u> $c_p \in M(X)$.

C3. If there is a commutative diagram of maps

$$\begin{array}{ccc} \tilde{X} & \xrightarrow{\tilde{h}} & \tilde{Y} \\ {\scriptstyle p_X}\downarrow & & \downarrow{\scriptstyle p_Y} \\ X & \xrightarrow{h} & Y \end{array}$$

with $\tilde{X}$, $\tilde{Y}$ two-sheeted covering spaces over X, Y and p_X, p_Y the corresponding covering maps, then

$$h^M(c_{p_Y}) = c_{p_X} .$$

C4. For a projective space P^n of dimension n the measure $M(P^n)$ is generated by the characteristic element $c_p \neq 0$ of the covering $p : S^n \to P^n$ with

$$c_p^{n+1} = 0 , \qquad c_p^n \neq 0 .$$

Clearly, a contradiction will follow from the properties C1 – C4. Hence, the theorem is proved in assuming the existence of such a measure, which may be taken e.g. to be the cohomology-ring over $\underline{Z}_2$ of a space.

Ex. D. Planar imbedding of linear graphs

A celebrated theorem of Kuratowski states, that for a linear graph G to be imbeddable in the plane it is necessary and sufficient, that G contains no subgraphs of one of the following two types:

Type 1. A complete graph of 5 vertices.

Type 2. A graph with two triads of vertices and all edges joining a pair of vertices, one from each triad.

The theorem of Kuratowski is, however, qualitative in character and is inpractical at all. The same holds for the criteria given by other authors, like Whitney and Mac Lane. On the other hand, we shall introduce some measure which permits to turn the qualitative theorem of Kuratowski into a quantitative one, so that the planarity of a linear graph can be verified by easy calculations.

For this let us construct for any space X the spaces $\tilde{X}_2$ and X_2, consisting of all ordered, respectively unordered pairs of distinct points of X. Take now the measure M on X to be the cohomology ring of X_2 over $\underline{Z}_2$:

$$M(X) = H^{*}_{\underline{Z}_2}(X_2) .$$

Let $\Phi'(X) \in H^{1}_{\underline{Z}_2}(X_2)$ be the ordinary characteristic class on $\underline{Z}_2$ of the two sheeted covering map $p : \tilde{X}_2 \to X_2$.

Set

$$\Phi^m(X) = \underbrace{\Phi'(X) \cup \dots \cup \Phi'(X)}_{m \text{ times}} \in H^m_{\underline{Z}_2}(X_2) .$$

Call $\Phi^2(X) \in M(X)$ the <u>characteristic element</u> c_p of the covering

$$p : \tilde{X}_2 \to X_2 .$$

Clearly, M and $c_p = \Phi^2$ verify properties D1 – D3, which are the same as C1 – C3 in Ex. C with, however, continuous maps $\tilde{h}$, h to be replaced by <u>imbeddings</u>. Besides, it is easy to see, that they verify the properties D4 – D5 below:

D4. For X = the plane R^2 we have

$$\Phi^2(R^2) = 0 .$$

D5. For X = any of the two Kuratowski graphs, we have

$$\Phi^2(X) \neq 0 .$$

It follows from these properties, that Kuratowski's theorem may be reformulated in the following way:

Theorem: A linear graph G is imbeddable in the plane if and only if

$$\Phi^2(G) = 0 .$$

The condition $\Phi^2(G) = 0$ may further be simplified by constructing a particular cocycle in $\Phi^2(G)$ as follows. Let G be connected, and take a maximal tree T of G through all vertices of G with one of the free vertices of T, say 0, to be designated as root of T. Distinguish the edges of G as internal or external ones, depending whether they belong to T or not. Now, imbed T arbitrarily in R^2, and then try to extend the whole of G to get an immersion f, so that no external edge can meet an internal one under f, except possibly at their extremities.

Now, consider any vertex v of G. To each pair of edges e_i, e_j of G incident with v, which are either internal or external but not lying on the path $0v$ in T from 0 to v, let us associate an unknown variable $x_{ij} = x_{ji}$ in $\underline{Z}_2$. For any free vertices a, b of G let the paths $a0$ and $b0$ in T from a and b to 0 respectively meet first at the vertex say v. If the edges incident with v on $a0$ and $b0$, but not on $v0$, be e_i and e_j, then we put $x_{ab} = x_{ij}$. By convention we also set $x_{aa} = 0$. For any two external edges $e_\alpha = \widehat{aa'}$ and $e_\beta = \widehat{bb'}$ joining free vertices a, a' and b, b' respectively, we set

$$x_{\alpha\beta} = x_{ab} + x_{ab'} + x_{a'b} + x_{a'b'} ,$$

the addition being carried out in $\underline{Z}_2$. Similarly, we set

$$f_{\alpha\beta} = I_f(e_\alpha, e_\beta) ,$$

viz. the intersection number mod 2 of $f(e_\alpha)$ and $f(e_\beta)$ with possible common extremities disregarded.

It is clear, that

$$x_{ab} = x_{ba} , \qquad x_{\alpha\beta} = x_{\beta\alpha}$$

and

$$f_{\alpha\beta} = f_{\beta\alpha} .$$

Now, form the system of linear equations with coefficients in $\underline{Z}_2$ and (α,β) running over all pairs of external edges of G :

$$(I)_f \qquad x_{\alpha\beta} = f_{\alpha\beta} .$$

Then, the Kuratowski theorem can further be restated as follows:

Theorem: A (connected) linear graph G is imbeddable in R^2 if and only if the linear system of equations $(I)_f$ obtained from any immersion f as described above is solvable in $\underline{Z}_2$.

Note

The computation-complexity of solving $(I)_f$ is utmost of order $O(n^b)$, where n is the number of edges in G . Moreover, as pointed out by Liu, each equation in $(I)_f$ will consist only of two terms in x_{ij} by suitably choosing the maximal tree T , so that the order of computation-complexity will be droped to $O(n^2)$. Furthermore, in case $(I)_f$ is solvable in $\underline{Z}_2$, we can adjoin new equations (this time quadratic) to $(I)_f$ and then modify the immersion f to a true imbedding from any solution (which necessarily exists) of the enlarged system. In fact, the totality of non-isomorphic imbeddings can be obtained in this way. For details cf. Wu [2] and Liu [1].

I.3 ADEQUACY OF MEASURES

The simple examples given in the preceeding section show, that the usefulness of a measure for applications relies much upon its nice properties in being easy to handle or easy to calculate. For this reason, we lay down for a given measure three general requirements, as some measuration of its usefulness or efficiency in applications, which will be expected to be observed.

Requirement 1: Invariance

The measure in question should be invariant in character to be meaningful. In algebraic topology, we are mostly interested in the topological and homotopic invariance for general topological spaces, the diffeomorphic invariance for differential manifolds, and the combinatorial invariance for complexes.

In this respect, the chain or cochain group- or ring-measures of complexes are not adequate, since they are not invariant with respect to any equivalence, meaningful in algebraic topology.

Requirement 2: Finiteness

The measure should be of a certain finite type, say possessing finite basis, in each degree, in case the measure in question is some graded algebraic structure.

In this respect, the singular chain or cochain-group or ring-measures of topological spaces are not adequate though they are topologically invariant by the very definition, so that Requirement 1 is fulfilled. The same holds for the graded-differential-algebraic measure of differential forms for differential manifolds, which is a diffeomorphic invariant, thus also meeting Requirement 1.

Requirement 3: Calculability or Constructibility

In order to explain this notion, let us first remark that in algebraic topology in particular we have to incessantly construct new spaces from given ones, as shown in the examples of last section, and then determine the algebraic measure in question of the newly constructed space in terms of those of original spaces. For example, among the earliest achievements in the development of algebraic topology we may particularly mention the theorems of Künneth and Mayer-Vietoris, which are just outcomes of trying to determine the betti-numbers of the space, constructed by union or product formation, from the betti-numbers of the given spaces. In the current formulation these theorems are expressed now in terms of homology or cohomology group measures and in the form of exact sequences. They will only give partial information about the measure in question of the new space constructed, the complete information is in fact impossible to get in this way. This is connected with the concept of calculability or constructibility to be formulated as follows.

Definition: Let X_α be a set of spaces in a certain category GEOM and CONS a certain geometric construction pertaining to GEOM, which produces some space X in GEOM from X_α. Then we say, that a measure Meas on GEOM is calculable with respect to the construction CONC, if Meas (X) is completely determined by means of some algebraic construction from Meas (X_α), together with the inherent interrelations of the latter ones, arising from their mutual geometrical relations and the geometrical construction CONS .

Non-calculability in the above sense occurs quite often in topology. The following are a few examples.

Ex. 1: The dimension-measure on the category of compact Hausdorff spaces is not calculabe with respect to the space-product construction. For, there are known examples, due to Pontrjagin, of such spaces X, Y of dimension, both equal to 2, while the product space $X \times Y$ has the dimension 3 instead of 4, which is usually the case. Thus, $\dim (X \times Y)$ cannot be determined from the knowledge of $\dim X$ and $\dim Y$ alone or dim is not calculable in this case.

Ex. 2: The Lebesgue measure (if defined) of point sets in a plane is calculable with respect to the union and intersection construction, but is non-calculable with respect to the derived-set construction.

Ex. 3: The integer measure of a knot K in the ordinary 3-space R^3, defined by the genus $g(K)$ or the signature $\sigma(K)$ of K, is calculable with respect to the connected-sum construction, since for any two knots K_1, K_2 we have (# means connected sum)

$$g(K_1 \# K_2) = g(K_1) + g(K_2) \quad \text{and}$$
$$\sigma(K_1 \# K_2) = \sigma(K_1) + \sigma(K_2) \ .$$

Ex. 4: For a pair of space X and a subspace Y consider the construction in removing Y from X to get the new space $X - Y$. Then the group-measure defined by the fundamental group π_1 is not calculable with respect to this removal construction, as it is seen from the following trivial example. For any knot K in $\underline{R}^3$, the morphism

$$\pi_1(K) \to \pi_1(\underline{R}^3)$$

relating the pair $(\underline{R}^3, K)$ is independent of K in $\underline{R}^3$, but $\pi_1(\underline{R}^3 - K)$ may be different for different K's.

The usual cohomology-group or ring measures are non-calculable even over the simple category of finite CW-complexes, with respect to such simple space-union or space-product constructions. In fact, we have the following table of calculability:

CONC \ Meas	$H^{\oplus}_{\underline{R}}$	$H^{\oplus}_{\underline{Z}}$	$H^{*}_{\underline{R}}$	$H^{*}_{\underline{Z}}$
SPACE-PRODUCT	Yes	Yes	Yes	No
SPACE-UNION	Yes	No	No	No

The answer Yes is simply due to the Künneth or Mayer-Vietoris exact sequence, which splits in these cases. The answer No may be seen from some simple examples given in Wu [10] which we reproduce below.

Ex. 5: Let K be the CW-complex consisting of a 0-cell e^0, two 2-cells e_1^2 and e_2^2 attached to e^0 to form two 2-spheres S_1^2 and S_2^2, a 3-cell e^3 with an attaching map $\dot{e}^3 \to S_2^2$ of degree 2, and a 4-cell e^4 with an attaching map $\dot{e}^4 \to S_1^2 \vee S_2^2$ which represents the Whitehead product. Let K' be the complex consisting of the same cells as K, with the only difference that e^4 will be attached to e^0 to form a 4-sphere S^4. Let L be a complex consisting of a 0-cell e_0^0, a 2-cell e_0^2 and a 3-cell e_0^3, such that $e_0^0 \vee e_0^2$ will be a 2-sphere S_0^2 and e_0^3 will be attached with a map $\dot{e}_0^3 \to S_0^2$ of degree 2.

Let e^3 also stand for the elementary cochain, taking value 1 on the cell e^3 and 0 on other cells. Similarly for the others. Then we see that

$H^2_{\underline{Z}}(K) \approx H^2_{\underline{Z}}(K') \approx \underline{Z}$ with generator $[e_1^2]$,

$H^3_{\underline{Z}}(K) \approx H^3_{\underline{Z}}(K') \approx \underline{Z}_2$ with generator $[e^3]$ of order 2,

$H^4_{\underline{Z}}(K) \approx H^4_{\underline{Z}}(K') \approx \underline{Z}$ with generator $[e^4]$,

$H^3_{\underline{Z}}(L) \approx \underline{Z}_2$ with generator $[e_0^3]$ of order 2.

All other groups $H^i_{\underline{Z}}$ $(i > 0)$ are 0. From the Künneth formulae we have $H^{\oplus}_{\underline{Z}}(K \times L) \approx H^{\oplus}_{\underline{Z}}(K' \times L)$, as listed in the following table

	$H^k_{\underline{Z}}(K \times L) \approx H^k_{\underline{Z}}(K' \times L)$	Generators
$k = 2$	$\underline{Z}$	$[e_1^2 \times e_0^0]$
$k = 3$	$\underline{Z}_2 + \underline{Z}_2$	$[e^3 \times e_0^0]$, $[e^0 \times e_0^3]$
$k = 4$	$\underline{Z}$	$[e^4 \times e_0^0]$
$k = 5$	$\underline{Z}_2 + \underline{Z}_2$	$[e_1^2 \times e_0^3]$, $[e_2^2 \times e_0^3 + e^3 \times e_0^2]$
$k = 6$	$\underline{Z}_2$	$[e^3 \times e_0^3]$
$k = 7$	$\underline{Z}_2$	$[e^4 \times e_0^3]$

Now, for the ring measure $H^*_{\underline{Z}}$ we see that both $H^*_{\underline{Z}}(K)$ and $H^*_{\underline{Z}}(K')$ have the trivial multiplicative structure. However, on the cochain level we have

$e_1^2 \cup e_2^2 = e^4$ in K, while

$e_1^2 \cup e_2^2 = 0$ in K'.

Consequently, $[e_1^2 \times e_0^0] \cup [e_2^2 \times e_0^3 + e_2^3 \times e_0^2] = 0$ in $K' \times L$ and $= [e^4 \times e_0^3] \neq 0$ in $K \times L$, so that

$$H_{\underline{Z}}^{*}(K \times L) \not\approx H_{\underline{Z}}^{*}(K' \times L) .$$

In conclusion, we have

Proposition 1: The $H_{\underline{Z}}^{*}$-measure is non calculable with respect to the space-product construction.

Ex. 6: Let K be the CW-complex consisting of a 0-cell e^0, a 2-cell e^2, two 3-cells e_1^3 and e_2^3, and a 4-cell e^4, with e^0, e^2 forming a 2-sphere S^2, and attaching maps $\dot{e}_1^3 \to S^2$, $\dot{e}_2^3 \to S^2$ of degrees 2 and 1 respectively. For the complex $e^0 \vee e^2 \vee e_1^3 \vee e_2^3$ we have by the Hurewicz theorem $\pi_3 \approx H_3^{\underline{Z}} \approx \underline{Z}$ with a generator corresponding to the cycle $e_1^3 - 2e_2^3$. Hence, we can attach e^4 by a map $\dot{e}^4 \to e^0 \vee e^2 \vee e_1^3 \vee e_2^3$, such that $\partial e^4 = 2e_1^3 - 4e_2^3$. Let K' be the complex consisting of a 0-cell e^0, a 2-cell e^2, two 3-cells e_1^3 and e_2^3 as in K, a 4-cell e^4 with an attaching map, such that $\partial e^4 = e_1^3 \mp 2e_2^3$, and another 3-cell e_0^3 and 4-cell e_0^4, such that e_0^3 and e^0 form a 3-sphere S_0^3, while e_0^4 is attached with a map $\dot{e}_0^4 \to S_0^3$ of degree 2. Let L be the subcomplex $e^0 \vee e^2 \vee e_1^3$ of K (and K'). Let C_L be the cone over L with a new vertex v and $\Delta = \Delta_L(K)$ (resp. $\Delta' = \Delta_L(K')$) be the union of K (resp. K') and C_L along the common subcomplex L. The Mayer-Vietoris sequence becomes now

$$0 \to \mathrm{Coker}^{k-1}\, i^H \to H_{\underline{Z}}^{k}(\Delta) \to \mathrm{Ker}^{k}\, i^H \to 0$$

in which

$$i^H : H_{\underline{Z}}^{\oplus}(K) \to H_{\underline{Z}}^{\oplus}(L)$$

is induced by the inclusion map $i : L \subset K$. Similarly for Δ'. Now, the only non-trivial homologies of K, K', L are:

$$H_{\underline{Z}}^{4}(K) \approx \underline{Z}_2 \quad \text{with generator} \quad [e^4] ,$$

$$H_{\underline{Z}}^{4}(K') \approx \underline{Z}_2 \quad \text{with generator} \quad [e_0^4] ,$$

$$H_{\underline{Z}}^{3}(L) \approx \underline{Z}_2 \quad \text{with generator} \quad [e_1^3] .$$

Hence, $H_{\underline{Z}}^{k}(\Delta) \approx H_{\underline{Z}}^{k}(\Delta')$ for $k \neq 4$, and the only case in doubt is for $k = 4$:

$$0 \to \underline{Z}_2 \to H_{\underline{Z}}^{4}(\Delta) \quad (\text{resp. } H_{\underline{Z}}^{4}(\Delta')) \to \underline{Z}_2 \to 0 .$$

It is easy to verify directly that

$$H^4_{\underline{Z}}(\Delta) \approx \underline{Z}_4 \qquad \text{with generator} \quad [e^4] \quad \text{of order } 4$$

$$H^4_{\underline{Z}}(\Delta') \approx \underline{Z}_2 \oplus \underline{Z}_2 \quad \text{with generators} \quad [e^4] \quad \text{and} \quad [e^4_0] \text{ , both of order } 2 .$$

Let us call the construction from (K, L) to Δ the <u>cone-construction</u> over the subcomplex L, Then we have

<u>Proposition 2</u>: The $H^{\oplus}_{\underline{Z}}$-measure is non-calculable with respect to the space-union construction. In fact, it is already non-calculable with respect to the cone-construction over a subspace.

<u>Ex. 7</u>: Let L be a 3-sphere and K (resp. K') be the mapping cylinder of L to a 2-sphere with a Hopf map (resp. a constant map). Let C_L be the cone over L and $\Delta = \Delta_L(K)$ (resp. $\Delta' = \Delta_L(K')$). Now,

$$i^* \;:\; H^*_{\underline{R}}(K) \to H^*_{\underline{R}}(L)$$

and

$$i'^* \;:\; H^*_{\underline{R}}(K') \to H^*_{\underline{R}}(L)$$

are the same for the inclusion map $i : L \subset K$ (resp. $i' : L \subset K'$). However, $H^*_{\underline{R}}(\Delta)$ and $H^*_{\underline{R}}(\Delta')$ are non-isomorphic in that Δ is homotopically $\underline{C}P$, while Δ' is $S^2 \vee S^4$. This proves the following

<u>Proposition 3</u>: The $H^*_{\underline{R}}$-measure is non-calculable with respect to the space-union construction. In fact, it is already non-calculable with respect to the cone-construction over a subspace.

It follows that the important measures, which we usually encounter in algebraic topology and which are very fertile in applications, do not meet the requirements of calculability. This causes much difficulties in the applications, and in algebraic topology various methods, notably methods of exact sequences and spectral sequences have been developed to turn around these difficulties. The problem naturally arises in finding a measure which will englobe the usual ones in algebraic topology and meets non-the-less all the three requirements listed above. If we forsake the finite or the torsion part of the usual homology and homotopy measures occuring in algebraic topology, then such a measure calculable with respect to the most important geometrical constructions does exist and was introduced and developed by D. Sullivan in the past decade. We call this measure due to Sullivan, the I^*-measure, whose study is the main concern of the present book. Whether the I^*-measure of Sullivan can further be extended to englobe also the finite or torsion part of the usual homology and homotopy measures in algebraic topology seems to be a very difficult, yet very important problem to settle. We hope, that the treatment given in the present book may serve to provide some clue to the solution of the problem.

I.4 GEOMETRICAL CATEGORY OF HOMOTOPIC - SIMPLICIAL SPACES

The purpose of the present book is a study of some measures originally due to D. Sullivan, from a constructive point of view, with an emphasis on its calculability with respect to various geometrical constructions which one encounters quite often in algebraic topology. In this respect the space-product and space-union formation are the simplest and also the most fundamental ones among the geometrical constructions to be studied. Thus, we have to choose the geometrical category of spaces to be one to which will pertain at least these space-product and space-union constructions. In view of this, the category of nilpotent spaces, on which is usually based the study of the measure due to Sullivan, is not adequate for our purposes, since it is not closed with respect to the cone-construction, a special case of the space-union construction. The category of CW-complexes is also not adequate, since it is not closed with respect to the space-product construction. As our fundamental category we shall choose instead the one of HCS-spaces defined below.

Definition: A space will be called a homotopic-simplicial space (abbr. HCS-space) if it is of the same homotopy type as the space of a countable connected simplicial complex in weak topology.

According to J.H.C. Whitehead, a countable connected CW-complex, and hence any space of the same homotopy type as such a complex, is a HCS-space. We add the remark, that in certain cases, such a CW-complex will not only be of the same homotopy type as, but will be itself a simplicial space after suitable simplicial subdivisions. For this, let us suppose that the CW-complex K verify the following inductive condition. For each attaching map f of a convex cell C^n to form an n-cell e^n of K, f will be a smooth map of the interior of each cell on the boundary of C^n to a certain cell on the boundary of e^n in the skeleton K^{n-1} with the highest rank, and e^n will then inherit the differential structure of C^n under f. Proceeding inductively on the dimension of the cells of K, we can then extend simplicial subdivisions, supposed already to exist on the (n-1)-skeleton of K and on the boundary of each C^n, with f simplicial to simplicial subdivisions of C^n and e^n to make f simplicial on the whole of C^n. In this way we get a simplicial subdivision of K so that the space of

K itself is a simplical space. Cf. for this also the work of Cairns [1] about triangulation of regular locally polyhedral spaces, of which the above CW-complexes are a special case.

We further remark that for π a countable abelian group there are $K(\pi,n)$ spaces, which are realizable as countable CW-complexes possessing simplicial subdivisions. In fact, according to Milgram [1] and Steenrod [1], we can take $K(\pi,n)$ for each $n \geq 0$ to be a topological abelian group as well as a CW-complex with cellular multiplication. The attaching maps of successive cells, as exhibited in the work of Milgram and Steenrod, are seen to be differential ones, so that we can simplicially subdivide in succession to get a simplical complex in the manner described above. Moreover, we have locally trivial fibrations

$$K(\pi,n-1) \underset{i}{\subset} E(\pi,n) \underset{g}{\to} K(\pi,n)$$

in which all spaces are CW-complexes with fibre $K(\pi,n-1)$ a subcomplex of the contractible space $E(\pi,n)$ and g a skeletal map. Using simplicial subdivisions in $K(\pi,n-1)$ and $K(\pi,n)$ as above we can carry the simplicial subdivisions of $K(\pi,n)$ over to $E(\pi,n)$, so that $K(\pi,n-1)$ will become a simplicial subcomplex of $E(\pi,n)$ and the projection g a simplicial map.

C h a p t e r   II

DGA AND MINIMAL MODEL

II.1 THE NOTION OF DGA AND ITS HOMOLOGY

Let $\underline{k}$ be a fixed field of characteristic 0, usually the field of rational numbers $\underline{Q}$. In this section, we gather the definitions of some well-known notions just for the sake of convenience. Recall first the definition of a DGA over $\underline{k}$ as a basic field for which $\underline{k}$ will often be understood when no misunderstanding can occur.

Definition: A graded module $\mathcal{M}$ over $\underline{k}$ (Abbr. GM/$\underline{k}$ or simply GM) is a module over $\underline{k}$ possessing a gradation verifying the following conditions:

$\mathcal{M}$ is the direct sum of linear spaces $\mathcal{M}_p$ over $\underline{k}$:

$$\mathcal{M} = \sum_{p \geq 0} \mathcal{M}_p .$$

The elements of $\mathcal{M}_p$ are said to be homogeneous of degree p, in notation:

$$\deg a = p , \qquad \text{for } a \in \mathcal{M}_p .$$

For our purposes we also add the following condition:

The field $\underline{k}$ itself is contained in $\mathcal{M}_0$, so that all elements of $\underline{k}$ are of degree 0.

Definition: A graded algebra $\mathcal{G}$ over $\underline{k}$ (abbr. GA/$\underline{k}$ or simply GA) is an associative algebra and at the same time a graded module over $\underline{k}$, so that the multiplication in $\mathcal{G}$ verifies the following conditions:

For homogeneous elements $a \in \mathcal{G}_p$ and $b \in \mathcal{G}_q$, ab is also homogeneous and belongs to $\mathcal{G}_{p+q}$. In other words,

$$\deg(ab) = \deg a + \deg b , \qquad \text{for } a \in \mathcal{G}_p , \quad b \in \mathcal{G}_q .$$

Moreover, the multiplication is anti-commutative, or

$$ab = (-1)^{pq} \cdot ba , \qquad \text{for } a \in \mathcal{G}_p , \quad b \in \mathcal{G}_q .$$

Definition: A differential graded algebra $\mathcal{A}$ over $\underline{k}$ (abbr. DGA/$\underline{k}$ or simply DGA) is a graded algebra, besides the gradation and the algebraic operations for an algebra also possessing a differentiation verifying the following conditions:

There is a k-linear operator $d_{\mathcal{A}}$ or simply d called the differential of $\mathcal{A}$, such that for any homogeneous elements a and b, da,db are also homogeneous and

$$\deg da = \deg a + 1 ,$$
$$d^2 a = 0 ,$$
$$d(ab) = da \cdot b + (-1)^{\deg a} \cdot a \cdot db ,$$
$$d/\underline{k} = 0 .$$

Notation: The subalgebra of DGA/k $\mathcal{A}$ generated by all elements of degree $\leq p$ will be denoted by $\mathcal{A}^{(p)}$. Thus,

$$\underline{k} \subset \mathcal{A}^{(0)} \subset \mathcal{A}^{(1)} \subset \ldots \subset \mathcal{A}^{(p)} \subset \ldots .$$

The ideal of $\mathcal{A}$ generated by all elements of degree > 0 will be denoted by $\mathcal{A}^+$. The ideal of $\mathcal{A}$ generated by all elements of the form $a_1 a_2$ with $a_1, a_2 \in \mathcal{A}^+$ will be denoted by $\mathcal{A}^+ \cdot \mathcal{A}^+$.

Definition: For the DGA $\mathcal{A}$ the elements in $\mathcal{A}^+$ are called positive, and the elements in $\mathcal{A}^+ \cdot \mathcal{A}^+$ are called decomposable.

Definition: Let $\mathcal{A}$ be a DGA/k. Then, the graded module (respectively the graded algebra) obtained from $\mathcal{A}$ by neglecting the multiplication and differentiation (respectively the differentiation) will be called the underlying graded module (respectively the underlying graded algebra) of $\mathcal{A}$ and will sometimes be denoted by Gmod $\mathcal{A}$ (respectively Galg $\mathcal{A}$).

Definition: The tensor product $\mathcal{A} \underset{\underline{k}}{\otimes} \mathcal{B}$ of two graded modules $\mathcal{A}$ and $\mathcal{B}$ over k is a graded module over k, which, as a k-module, is the usual tensor product of modules $\mathcal{A}$ and $\mathcal{B}$, while the gradation is such, that for $a \in \mathcal{A}$, $b \in \mathcal{B}$ homogeneous, $a \otimes b \in \mathcal{A} \underset{\underline{k}}{\otimes} \mathcal{B}$ is also homogeneous, and

$$\deg (a \otimes b) = \deg a + \deg b .$$

Definition: The tensor product $\mathcal{A} \underset{\underline{k}}{\otimes} \mathcal{B}$ of the graded algebras $\mathcal{A}$ and $\mathcal{B}$ over k is the graded algebra over k which, as a graded module, is the previously defined tensor product of $\mathcal{A}$ and $\mathcal{B}$, while as an algebra the gradation and multiplication verify the further condition below:

For any homogeneous element a, $a' \in \mathcal{A}$ and b, $b' \in \mathcal{B}$,

$$(a \otimes b)(a' \otimes b') = (-1)^{\deg b \cdot \deg a'} \cdot aa' \otimes bb' .$$

Definition: The tensor product $\mathcal{A} \otimes_{\underline{k}} \mathcal{B}$ of two DGA/$\underline{k}$'s $\mathcal{A}$ and $\mathcal{B}$ is the DGA/$\underline{k}$ which, as a graded algebra, is the tensor product of the underlying graded algebras $\mathcal{A}$ and $\mathcal{B}$, as previously defined, while the differentiation verifies the further condition below:

$$d(a \otimes b) = da \otimes b + (-1)^{\deg a} \cdot a \otimes db$$

for any homogeneous element $a \in \mathcal{A}$, $b \in \mathcal{B}$.

Notations: We shall simply write $\mathcal{A} \otimes \mathcal{B}$ instead of $\mathcal{A} \otimes_{\underline{k}} \mathcal{B}$, whenever no misunderstanding can occur. Furthermore, for any $a \in \mathcal{A}$ and $b \in \mathcal{B}$, instead of $a \otimes 1$ or $1 \otimes b$ in $\mathcal{A} \otimes \mathcal{B}$, we shall simply write a or b in short. Accordingly, the element $a \otimes b$ will simply be written ab and also $(-1)^{\deg a \cdot \deg b}$ ba, if no misunderstanding can occur. Remark also that $\mathcal{A} \otimes \mathcal{B}$ is canonically isomorphic to $\mathcal{B} \otimes \mathcal{A}$ under the natural correspondence $a \otimes b \leftrightarrow (-1)^{\deg a \cdot \operatorname{deb} b}$ $b \otimes a$, so that sometimes we shall write $\mathcal{A} \otimes \mathcal{B}$ alternatively as $\mathcal{B} \otimes \mathcal{A}$, when no misunderstanding can occur.

Definition: For a DGA/$\underline{k}$ $\mathcal{A}$: the morphism $\mathcal{A} \to \mathcal{A}_0$, which is identity on $\mathcal{A}_0$ and 0 on $\mathcal{A}^+$, is called the augmentation of $\mathcal{A}$. The morphism will be denoted by $\varepsilon_{\mathcal{A}}$ or simply ε.

Definition: A GA/$\underline{k}$ $\mathcal{A}$ is said to be free if the multiplication in $\mathcal{A}$ obeys no rules besides the usual ones, including the anti-commutativity rule:

$$ab = (-1)^{pq} \cdot ba \qquad \text{for } a \in \mathcal{A}_p, \quad b \in \mathcal{A}_q .$$

It is then freely generated by a system of free generators, say $\{z_\gamma\}$. The free algebra will then be denoted by $\mathcal{A} = \operatorname{Free}(z_\gamma)$. A free GA/$\underline{k}$ on free generators x_α (resp. y_β), all of odd (resp. even) degree, is called an exterior algebra on $\underline{k}$ (resp. polynomial algebra on $\underline{k}$) to be denoted by $\operatorname{Extr}(x_\alpha)$ (resp. $\operatorname{Polym}(y_\beta)$).

It is clear that any free GA/$\underline{k}$ is the tensor product of an exterior algebra and a polynomial algebra. Now, the following definition will be seen to be important in Chapter V.

Definition: A Cartan algebra $\mathcal{A}$ is a DGA/$\underline{k}$ which is free as a graded algebra, so that

$$\mathcal{A} = \operatorname{Polym}(y_\beta) \otimes \operatorname{Extr}(x_\alpha) ,$$

while the differentiation satisfies the following conditions:

$dx_\alpha \in \text{Polym}(y_\beta)$,
$dy_\beta = 0$.

The differential d, sending elements (x_α) to $\text{Polym}(y_\beta)$ is said to be twisted.

Definition: For a DGA/k A a (homogeneous) element z (of degree p) is called a cycle of degree p) if $dz = 0$ and a boundary if it is of the form $z = da$ for some element a of A. The homogeneous cycles (resp. boundaries) of same degree p form naturally an abelian group to be called the group of cycles resp. boundaries of A and will be denoted by $\text{Cyc}_p A$ resp. $\text{Bdry}_p A$ with $\text{Bdry}_p A$ a subgroup of $\text{Cyc}_p A$. The quotient group

$$H_p(A) = \text{Cyc}_p A / \text{Bdry}_p A$$

is then called the p-th homology group of A. The elements of $H_p(A)$ are called homology classes of degree p. The cycles z_i $(i = 1,2)$ of same degree p are said to be homologous in A, denoted as $z_1 \sim z_2$ in A if $z_1 - z_2 \in \text{Bdry}_p A$. The multiplication in A will induce naturally one in the direct sum

$$H(A) = \sum_{p \geq 0} H_p(A)$$

with

$$H_p(A) \cdot H_q(A) \subset H_{p+q}(A) .$$

With this multiplication $H(A)$ becomes GA/k and also a DGA/k, if we take the differential in $H(A)$ to be $d = 0$. This GA/k resp. DGA/k will then be called the homology algebra resp. the homology DGA of the given DGA A. The passage H from A to $H(A)$ will be called the homology operator of A.

Definition: A DGA/k A is called connected if $H_0(A) \approx k$, and is simply-connected if it is connected as well as $H_1(A) = 0$. It is said to be of finite type if for each i, $H_i(A)$ is a vector space of finite dimension over k.

In the later sections of this chapter we shall consider DGA's over a fixed k, and so k will be omitted throughout in these sections.

Remark: In the next chapter we have to consider non-commutative DGA , which are the same as DGA with, however, the law of anti-commutative multiplications $ab = (-1)^{\deg a \cdot \deg b} \cdot ba$ for homogeneous a , b removed. The various notions about DGA extends to this non-commutative case in a natural manner. In particular, we may speak of DGA-morphisms, homologies, etc. just in the same way, as far as there is no anticommutativity of multiplications involved.

II.2 DGA-MORPHISMS AND HOMOTOPY OF DGA's

Definition: A DGA-morphism

$$f : \mathcal{A} \to \mathcal{B}$$

of a DGA $\mathcal{A}$ to a DGA $\mathcal{B}$ is a morphism of algebra from $\mathcal{A}$ to $\mathcal{B}$, which preserves gradation and differentiation, i.e.

$$\deg f(a) = \deg a$$

and

$$df(a) = f(da)$$

for any homogeneous element a of $\mathcal{A}$.

A DGA-morphism

$$f : \mathcal{A} \to \mathcal{B}$$

of a DGA $\mathcal{A}$ to DGA $\mathcal{B}$ is such that

$$f(\mathrm{Cyc}_p \mathcal{A}) \subset \mathrm{Cyc}_p \mathcal{B},$$
$$f(\mathrm{Bdry}_p \mathcal{A}) \subset \mathrm{Bdry}_p \mathcal{B} .$$

Hence, it will naturally induce morphisms

$$f_H : H(\mathcal{A}) \to H(\mathcal{B}) \ , \quad \text{and}$$
$$f_{H_i} \ \text{ or } \ f_H : H_i(\mathcal{A}) \to H_i(\mathcal{B}) \ .$$

The first one is an algebra-morphism resp. a DGA-morphism, if $H(\mathcal{A})$, $H(\mathcal{B})$ are considered GA's resp. DGA's . The notation f_H with H in the subscript indicates the covariant character of the induced morphism.

Definition: The morphism $f_H : H(\mathcal{A}) \to H(\mathcal{B})$ will be called the H-morphism induced by the DGA-morphism $f : \mathcal{A} \to \mathcal{B}$. The DGA-morphism f will be said to be an H-isomorphism (or one in degree i) if f_H induced is an isomorphism $f_H : H(\mathcal{A}) \approx H(\mathcal{B})$ (or one in degree i, $f_{H_i} : H_i(\mathcal{A}) \approx H_i(\mathcal{B})$).

Definition: The trivial DGA in t, denoted by $Tr(t)$, is the DGA/$\underline{k}$, generated freely by two elements t and t^+ as algebra with gradation and differentiation, given by

$$\deg t = 0, \qquad \deg t^+ = 1,$$
$$dt = t^+, \qquad dt^+ = 0.$$

According to § II.1, the trivial DGA $Tr(t)$ is somewhat alike to a Cartan algebra, which, as an algebra, is just the tensor product of a polynomial algebra $Polym(t)$ in t and an exterior algebra in t^+:

$$Tr(t) = Polym(t) \otimes Extr(t^+),$$

while the differential in $Tr(t)$ is a twisted one, given by $dt = t^+$. Henceforth, we shall write dt instead of t^+, so that as graded algebra,

$$Tr(t) = Polym(t) \otimes Extr(dt) = Free(t, dt).$$

Any element z of $Tr(t)$ can thus be expressed in the form

$$z = p(t) + q(t)dt,$$

where $p(t)$, $q(t) \in \underline{k}[t]$ are polynomials in t with coefficients in the field $\underline{k}$.

Definition: For any DGA $\mathcal{A}$ the tensor product $\mathcal{A} \otimes Tr(t)$ will be called the trivial extension of $\mathcal{A}$ by t and will be denoted

$$\mathcal{A}^t = \mathcal{A} \otimes Tr(t).$$

For $i = 0, 1$, the DGA-morphism

$$\tau_i : \mathcal{A}^t \to \mathcal{A}$$

defined by

$$\tau_i(a) = a\,, \qquad a \in \mathcal{A}\,,$$
$$\tau_i(t) = i\,, \qquad \tau_i(dt) = 0$$

will be called the i-projection of $\mathcal{A}^t$.

Since $\underline{k}$ is a field of characteristic 0 , for any polynomial $p(t) \in \underline{k}[t]$ we can speak of its derivative and its integral to be defined formally in the usual manner with the usual notation $\frac{dp}{dt}$, $\int_0^t p(t)\,dt$, $\int_0^1 p(t)\,dt$, etc. Thus, for $k_i \in \underline{k}$, and

$$p(t) = k_0 + k_1 t + \ldots + k_n t^n \in \underline{k}[t]\,,$$

we have formally by definition

$$\frac{dp}{dt} = k_1 + 2k_2 t + \ldots + nk_n t^{n-1}\,,$$
$$\int_0^t p(t)\,dt = k_0 t + k_1 \cdot \frac{t}{2} + \ldots + k_n \cdot \frac{t^{n+1}}{n+1}\,,$$
$$\int_0^1 p(t)\,dt = k_0 + \frac{k_1}{2} + \ldots + \frac{k_n}{n+1}\,.$$

Now, any element a of $\mathcal{A}^t$ can be written in the form

$$a = a' + dt \cdot a''$$

with a' , $a'' \in \mathcal{A} \otimes \mathrm{Polym}(t)$, say

$$a' = \sum_{i \geq 0} a'_i \cdot \frac{t^i}{i!}\,, \qquad a'_i \in \mathcal{A}\,,$$
$$a'' = \sum_{i \geq 0} a''_i \cdot \frac{t^i}{i!}\,, \qquad a''_i \in \mathcal{A}\,.$$

We then extend the definition of differentiation and integration to a' and a'' , viz. by

$$\frac{da'}{dt} = \sum_{i \geq 0} a'_i \cdot \frac{t^{i-1}}{(i-1)!}\,,$$
$$\int_0^t a'dt = \sum_{i \geq 0} a'_i \cdot \frac{t^{i+1}}{(i+1)!}\,,$$
$$\int_0^1 a'dt = \sum_{i \geq 0} \frac{a'_i}{(i+1)!}\,.$$

We also extend the differential d in $\mathcal{A}$ to d' in $\mathcal{A} \otimes \mathrm{Polym}(t)$ simply as $d \otimes 1$, so that

$$d'a' = \sum_{i \geq 0} da'_i \cdot \frac{t^i}{i!} = da' - dt \cdot \frac{da'}{dt}\,.$$

Definition: The I_t-operator (of A. Weil)

$$I_t : A^t \to A^t$$

is the k-module morphism given by

$$I_t(a' + dt \cdot a'') = \int_0^t a''\,dt .$$

We also define I_i by

$$I_i = \tau_i I_t : A^t \to A .$$

We also sometimes write I instead of I_1 . Then, $I_0 = 0$ and

$$I(a' + dt \cdot a'') = \int_0^1 a''\,dt .$$

Lemma 1: $I_t d + dI_t = 1 - \tau_0 : A^t \to A^t$.

Proof: Let $a = a' + dt \cdot a''$ as above. Then,

$$dI_t a = d\int_0^t a''\,dt = d\,\Sigma\, a''_i \cdot \frac{t^{i+1}}{(i+1)!}$$

$$= dt \cdot \Sigma\, a''_i \cdot \frac{t^i}{i!} + \Sigma\, da''_i \cdot \frac{t^{i+1}}{(i+1)!}$$

$$= dt \cdot a'' + \int_0^t (d'a'')\,dt .$$

On the other hand,

$$da = d'a' + dt\,(\frac{da'}{da} - d'a'') .$$

Hence

$$I_t da + dI_t a = dt \cdot a'' + \int_0^t \frac{da'}{dt}\,dt = dt \cdot a'' + a' - \tau_0 a' = a - \tau_0 a$$

as to be proved.

Lemma 2: $Id + dI = \tau_1 - \tau_0 : A^t \to A$.

Proof: By definition and using Lemma 1, we have

$$Id + dI = \tau_1 I_t d + d(\tau_1 I_t) = \tau_1(I_t d + dI_t) = \tau_1(1 - \tau_0) = \tau_1 - \tau_0 ,$$

as to be proved.

Definition: Given two DGA's $\mathcal{A},\mathcal{B}$ and two DGA-morphisms f_0, f_1 of $\mathcal{A}$ to $\mathcal{B}$, f_0 will be said to be homotopic to f_1, if there exists a DGA-morphism

$$H : \mathcal{A} \to \mathcal{B}^t ,$$

such that the following diagram of DGA's and DGA-morphisms is commutative:

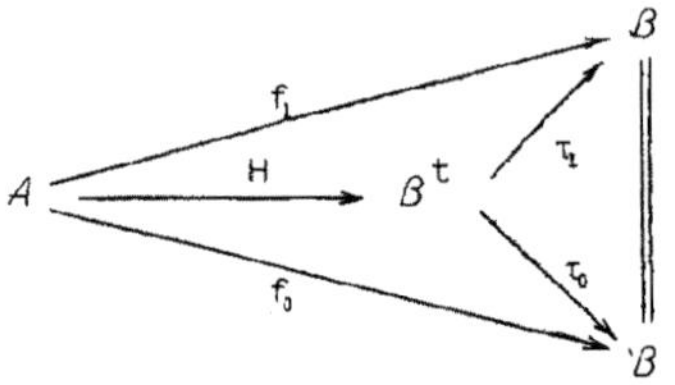

We then use the notation:

$$f_0 \simeq f_1 : \mathcal{A} \to \mathcal{B} .$$

Proposition 1: For given DGA's $\mathcal{A}$ and $\mathcal{B}$ the homotopy of DGA-morphism of $\mathcal{A}$ to $\mathcal{B}$ is an equivalence relation.

Proof: Only transitivity requires proof. To see this, let $H_0 : \mathcal{A} \to \mathcal{B}^t$ realize a homotopy $f_0 \simeq f_1 : \mathcal{A} \to \mathcal{B}$ and $H_1 : \mathcal{A} \to \mathcal{B}^t$ a homotopy $f_1 \simeq f_2 : \mathcal{A} \to \mathcal{B}$. Then

$$H = H_0 + H_1 - \tau_0 H_1$$

clearly realizes a homotopy $f_0 \simeq f_2$, since $\tau_1 H_0 = \tau_0 H_1 = f_1$ and $\tau_i \tau_0 = \tau_0$, $i = 0, 1$.

Proposition 2: Homotopic DGA-morphisms induce identical H-morphisms, i.e. if

$$f_0 \simeq f_1 : \mathcal{A} \to \mathcal{B} ,$$

then

$$f_{0_H} = f_{1_H} : H(\mathcal{A}) \to H(\mathcal{B}) .$$

Proof: Let z be any (homogeneous) cycle of $\mathcal{A}$. As H is a DGA-morphism

$$dHz = Hdz = 0 .$$

By the commutativity of the diagram and Lemma 2 above we therefore get

$$\begin{aligned} f_1 z - f_0 z &= \tau_1 Hz - \tau_0 Hz \\ &= (dI + Id)Hz \\ &= dIHz \\ &\sim 0 \quad \text{in } \mathcal{B}. \end{aligned}$$

Hence, $f_0 z \sim f_1 z$ in $\mathcal{B}$ or $f_{0_H} = f_{1_H}$, as it will be proved.

II.3 MINIMAL MODEL OF A DGA-EXISTENCE

Definition: A DGA/$\underline{k}$ $\mathcal{M}$ is called minimal if the following conditions are verified:

1°. Freeness: It is free as a graded algebra.

2°. Connectedness: It is connected, i.e., $H_0(\mathcal{M}) \approx \underline{k}$, or what is the same, $\mathcal{M}_0 \approx \underline{k}$.

3°. Decomposability: For any element $x \in \mathcal{M}^+$, the differential dx is decomposable, i.e.

$$dx \in \mathcal{M}^+ \cdot \mathcal{M}^+ .$$

4°. Each $\mathcal{M}^{(n)}$ for $n \geq 1$ is the union of a finite or infinite increasing sequence of DGA's

$$\mathcal{M}_0^{(n)} \subset \mathcal{M}_1^{(n)} \subset \ldots \subset \mathcal{M}_m^{(n)} \subset \ldots$$

with

$$\mathcal{M}_0^{(1)} \approx H_1(\mathcal{M}),$$
$$d\mathcal{M}_m^{(n)} \subset \mathcal{M}_{m-1}^{(n)} \quad \text{for } m \geq 0,$$

where we have put for convenience

$$\mathcal{M}_{-1}^{(n)} = \mathcal{M}^{(n-1)} .$$

Notation: The $\underline{k}$-vector space spanned by free generators of degree n in $\mathcal{M}$ will be denoted by $\text{Vect}_n\, \mathcal{M}$.

Theorem: For any connected DGA/$\underline{k}$ $\mathcal{A}$ there is a minimal DGA/$\underline{k}$ $\mathcal{M}$ and a DGA-morphism

$$\rho : \mathcal{M} \to \mathcal{A}$$

which is an H-isomorphism, i.e.

$$\rho_H : H(\mathcal{M}) \approx H(\mathcal{A}) .$$

Idea of the proof: We simply put

$$\mathcal{M}^{(0)} = \mathcal{M}_0 = \underline{k}$$

and then enlarge $\mathcal{M}^{(0)}$ successively to $\mathcal{M}^{(1)}, \ldots, \mathcal{M}^{(n)}, \ldots$ by adjoining new free generators of degree $n+1$ to $\mathcal{M}^{(n)}$ to form $\mathcal{M}^{(n+1)}$. At the same time we enlarge successively DGA-morphisms

$$\rho^{(n)} : \mathcal{M}^{(n)} \to \mathcal{A}$$

to $\rho^{(n+1)} : \mathcal{M}^{(n+1)} \to \mathcal{A}$ with restriction

$$\rho^{(n+1)} \mid \mathcal{M}^{(n)} = \rho^{(n)} .$$

Now, to each $\rho^{(n)}$ we have exact sequences

$$0 \to \operatorname{Ker} \rho^{(n)}_{H_j} \to H_j(\mathcal{M}^{(n)}) \xrightarrow{\rho^{(n)}_{H_j}} H_j(\mathcal{A}) \to \operatorname{Coker} \rho^{(n)}_{H_j} \to 0 .$$

The principle of the construction of $\mathcal{M}^{(n+1)}$ is to introduce new generators to $\mathcal{M}^{(n)}$ in order to kill $\operatorname{Ker} \rho^{(n)}_{H_n}$ and $\operatorname{Coker} \rho^{(n)}_{H_{n+1}}$, so that the following induction hypothesis will be observed:

Ind. Hyp. $1^{(n)}$.

$$\rho^{(n)}_{H_j} : H_j(\mathcal{M}^{(n)}) \to H_j(\mathcal{A})$$

are isomorphisms for $j \leq n$.

Ind. Hyp. $2^{(n)}$.

$$\rho^{(n)}_{H_{n+1}} : H_{n+1}(\mathcal{M}^{(n)}) \to H_{n+1}(\mathcal{A})$$

is a monomorphism.

Proof of Theorem: Let us begin by setting

$$\mathcal{M}^{(0)} = \mathcal{M}_0 = \underline{k} .$$

Then, the induction hypothesis $1^{(0)}$ and $2^{(0)}$ are trivially satisfied. Suppose now, the inductive construction has been done in step n so that $\mathcal{M}^{(n)}$ and

$$\rho^{(n)} : \mathcal{M}^{(n)} \to \mathcal{A}$$

have been defined verifying the inductive hypothesis $1^{(n)}$ and $2^{(n)}$. Consider now the short exact sequence

$$H_{n+1}(\mathcal{M}^{(n)}) \xrightarrow{\rho^{(n)}_{H_{n+1}}} H_{n+1}(\mathcal{A}) \to \text{Coker}\, \rho^{(n)}_{H_{n+1}} \to 0 .$$

Take a $\underline{k}$-basis $\{X_j^{(n+1)} \bmod \operatorname{Im} \rho^{(n)}_{H_{n+1}}\}$ of Coker $\rho^{(n)}_{H_{n+1}}$ and a cycle $x_j^{(n+1)}$ in each class $X_j^{(n+1)} \in H_{n+1}(\mathcal{A})$.

Corresponding to each $x_j^{(n+1)}$ adjoin a new free generator $\xi_j^{(n+1)}$ to $\mathcal{M}^{(n)}$ with

$$\deg \xi_j^{(n+1)} = n+1 ,$$
$$d\ \xi_j^{(n+1)} = 0 .$$

The new DGA thus formed will be denoted by

$$\mathcal{M}_0^{(n+1)} = \mathcal{M}^{(n)} \otimes \text{Free}\,(\xi_j^{(n+1)}) .$$

Extend also $\rho^{(n)}$ to a DGA-morphism

$$\rho_0^{(n+1)} : \mathcal{M}_0^{(n+1)} \to \mathcal{A}$$

by setting

$$\rho_0^{(n+1)} : (\xi_j^{(n+1)}) = x_j^{(n+1)}$$

while

$$\rho_0^{(n+1)} \mid \mathcal{M}^{(n)} = \rho^{(n)} .$$

From $\mathcal{M}_0^{(n+1)}$ we shall now adjoin successively new free generators to get an increasing sequence of DGA's

$$\mathcal{M}_0^{(n+1)} \subset \mathcal{M}_1^{(n+1)} \subset \ldots \subset \mathcal{M}_m^{(n+1)} \subset \ldots$$

and define also DGA-morphisms

$$\rho_m^{(n+1)} : \mathcal{M}_m^{(n+1)} \to \mathcal{A}$$

with

$$\rho_m^{(n+1)} \mid \mathcal{M}_{m-1}^{(n+1)} = \rho_{m-1}^{(n+1)}$$

as follows.

Suppose that $\mathcal{M}_m^{(n+1)}$ and

$$\rho_m^{(n+1)} : \mathcal{M}_m^{(n+1)} \to \mathcal{A}$$

have already been defined. Consider the short exact sequence

$$0 \to \operatorname{Ker} \rho^{(n+1)}_{m,H_{n+2}} \to H_{n+2}(\mathcal{M}^{(n+1)}_m) \xrightarrow{\rho^{(n+1)}_{m,H_{n+2}}} H_{n+2}(\mathcal{A}) .$$

The construction will be stopped at $\mathcal{M}^{(n+1)}_m$ if $\operatorname{Ker} \rho^{(n+1)}_{m,H_{n+2}} = 0$.

In this case we shall set for all $k \geq 0$. $\mathcal{M}^{(n+1)}_{m+k} = \mathcal{M}^{(n+1)}_m$ and define $\rho^{(n+1)}_{m+k} : \mathcal{M}^{(n+1)}_{m+k} \to \mathcal{A}$ by $\rho^{(n+1)}_{m+k} = \rho^{(n+1)}_m$. Otherwise take a $\underline{k}$-basis $\{\Gamma^{(n+1)}_{mj}\}$ of $\operatorname{Ker} \rho^{(n+1)}_{m,H_{n+1}}$ and a cycle $\gamma^{(n+1)}_{mj} \in \mathcal{M}^{(n+\ )}_m$ in each $\Gamma^{(n+1)}_{mj}$. Then

$$\rho^{(n+1)}_m \gamma^{(n+1)}_{mj} \sim 0 \quad \text{in } \mathcal{A}$$

so that

$$\rho^{(n+1)}_m \gamma^{(n+1)}_{mj} = dc^{(n+1)}_{mj}$$

for some $c^{(n+1)}_{mj}$ in $\mathcal{A}$. Corresponding to $c^{(n+1)}_{mj}$ adjoin new free generators $\zeta^{(n+1)}_{mj}$ to $\mathcal{M}^{(n+1)}_m$ with

$$\deg \zeta^{(n+1)}_{mj} = n + 1 ,$$
$$d\,\zeta^{(n+1)}_{mj} = \gamma^{(n+1)}_{mj} \in \mathcal{M}^{(n+1)}_m .$$

The new DGA thus formed will be denotec by

$$\mathcal{M}^{(n+1)}_{m+1} = \mathcal{M}^{(n+1)}_m \otimes \text{Free}\,(\zeta^{(n+1)}_{mj})$$

with degree and differential d as above.

As

$$\rho^{(n+1)}_m \, d\zeta^{(n+1)}_{mj} = \rho^{(n+1)}_m \gamma^{(n+1)}_{mj} = dc^{(n+1)}_{mj} ,$$

we can extend also $\rho^{(n+1)}_m$ already defined to a DGA-morphism

$$\rho^{(n+1)}_{m+1} : \mathcal{M}^{(n+1)}_{m+1} \to \mathcal{A}$$

by setting

$$\rho^{(n+1)}_{m+1} (\zeta^{(n+1)}_{mj}) = c^{(n+1)}_{mj} ,$$
$$\rho^{(n+1)}_{m+1} \mid \mathcal{M}^{(n+1)}_m = \rho^{(n+1)}_m .$$

Now, define the DGA $\mathcal{M}^{(n+1)}$ as the union of all $\mathcal{M}^{(n+1)}_m$ thus obtained and also the DGA-morphism

$$\rho^{(n+1)} : \mathcal{M}^{(n+1)} \to \mathcal{A}$$

by

$$\rho^{(n+1)} \mid \mathcal{M}_m^{(n+1)} = \rho_m^{(n+1)}$$

for all m. It is easy to see from the construction that the inductive hypothesis $1^{(n+1)}$ and $2^{(n+1)}$ are both verified in this $(n+1)$-th step. Moreover, from mere degree considerations we see that each $Y_{mj}^{(n+1)}$ is decomposable, so that each $\mathcal{M}_m^{(n+1)}$ and also $\mathcal{M}^{(n+1)}$ are minimal DGA's.

The procedure can be done indefinitely if necessary, and we get thus an increasing sequence of minimal DGA's

$$\mathcal{M}^{(0)} \subset \mathcal{M}^{(1)} \subset \ldots \subset \mathcal{M}^{(n)} \subset \ldots$$

and a sequence of DGA-morphism

$$\rho^{(n)} : \mathcal{M}^{(n)} \to \mathcal{A}$$

with

$$\rho^{(n+1)} \mid \mathcal{M}^{(n)} = \rho^{(n)}$$

for all n, the inductive hypothesis being verified for each n.

Define now the DGA $\mathcal{M}$ as the union of all $\mathcal{M}^{(n)}$ and also the DGA-morphism

$$\rho : \mathcal{M} \to \mathcal{A}$$

by setting

$$\rho \mid \mathcal{M}^{(n)} = \rho^{(n)}$$

for all n. Then it is clear that $\mathcal{M}$ is a minimal DGA with ρ an H-isomorphism as to be asserted. q.e.d.

Definition: The minimal DGA/$\underline{k}$ $\mathcal{M}$ whose existence is asserted in the theorem is called a minimal model of $\mathcal{A}$, and the DGA-morphism $\rho : \mathcal{M} \to \mathcal{A}$ inducing an H-isomorphism is called a minimal morphism associated to $\mathcal{M}$ or $\mathcal{A}$.

In the next section we shall prove that $\mathcal{M}$ is unique up to DGA-isomorphism though ρ is not. Accordingly we shall adopt the notation

$$\mathcal{M} = \min \mathcal{A} .$$

II.4 MINIMAL MODEL OF A DGA–UNIQUENESS

For any DGA $\mathcal{A}$ we have proved, in the last section, the existence of a minimal model $\mathcal{M}$ of $\mathcal{A}$ as well as an associated minimal morphism

$$\rho : \mathcal{M} \to \mathcal{A}$$

which induces an H-isomorphism. The present section will prove that such a minimal model is unique up to DGA-isomorphism, while the associated minimal morphisms are unique only up to homotopy in the sense that for any two such associated minimal morphisms ρ, ρ' we have a DGA-isomorphism $i : \mathcal{M} \approx \mathcal{M}'$, such that

$$\rho' i \simeq \rho' : \mathcal{M} \to \mathcal{A} .$$

To begin with, let us first prove the following

Lemma 1: Let $\mathcal{M}, \bar{\mathcal{M}}$ be both, minimal DGA's and

$$\varphi : \mathcal{M} \to \bar{\mathcal{M}}$$

be a DGA-morphism which is identity on $\mathcal{M}_0 = \bar{\mathcal{M}}_0 = \underline{k}$ and induces an H-isomorphism, viz.

$$\varphi_H : H(\mathcal{M}) \approx H(\bar{\mathcal{M}}) .$$

Then, φ is itself an isomorphism.

Proof: Denote, as in §II 3, $\mathcal{M}$ as the union of sub-DGA's $\mathcal{M}^{(n)}$ and each $\mathcal{M}^{(n)}$ as the union of sub-DGA's $\mathcal{M}^{(n)}_m$ with similar notations for $\bar{\mathcal{M}}$, where we have put $\mathcal{M}^{(n+1)}_{-1} = \mathcal{M}^{(n)}$ for convenience. Then φ, when restricted to $\mathcal{M}^{(n)}$ and $\mathcal{M}^{(n)}_m$, will give DGA-morphisms

$$\varphi^{(n)} : \mathcal{M}^{(n)} \to \bar{\mathcal{M}}^{(n)} \quad \text{and}$$

$$\varphi^{(n)}_m : \mathcal{M}^{(n)}_m \to \bar{\mathcal{M}}^{(n)} .$$

It is clear that

$$\varphi^{(0)} : \mathcal{M}^{(0)} \approx \bar{\mathcal{M}}^{(0)} \quad (\approx \underline{k})$$

and

$$\varphi^{(1)}_0 : \mathcal{M}^{(1)}_0 \approx \bar{\mathcal{M}}^{(1)}_0$$

too, since

$$\mathcal{M}^{(1)}_0 \approx H_1(\mathcal{M}) , \quad \bar{\mathcal{M}}^{(1)}_0 \approx H_1(\bar{\mathcal{M}})$$

and

$$\varphi_{H_1} : H_1(\mathcal{M}) \approx H_1(\bar{\mathcal{M}}) .$$

Suppose now $\varphi^{(n+1)}_{m-1}$ $(m \geq 0)$ has already been proved to be a monomorphism. We shall prove that $\varphi^{(n+1)}_{m}$ is also a monomorphism.

To see this, let x be any element of degree $n+1$ in $\mathcal{M}^{(n+1)}_{m}$ but not in $\mathcal{M}^{(n+1)}_{m-1}$ with $\varphi x = 0$. As $dx \in \mathcal{M}^{(n+1)}_{m-1}$, $\varphi^{(n+1)}_{m-1} dx \in \mathcal{M}^{(n+1)}$ is well defined. Now, $\varphi^{(n+1)}_{m-1} dx = \varphi dx = d\varphi x = 0$. As $\varphi^{(n+1)}_{m-1}$ is a monomorphism, it follows that $dx = 0$ or x is a cycle. As $\varphi x = 0$ is trivially ~ 0 in $\bar{\mathcal{M}}$, so, φ_H being an isomorphism, x itself is ~ 0 in $\mathcal{M}$. Consequently $x = da$ for some $a \in \mathcal{M}$. As $\deg a = n$, a is necessarily in $\mathcal{M}^{(n)} \subset \mathcal{M}^{(n+1)}_{m-1}$, so that $\varphi^{(n+1)}_{m-1} a$ is well-defined. Now $\varphi^{(n+1)}_{m-1} da =$ $= \varphi da = \varphi x = 0$. As $\varphi^{(n+1)}_{m-1}$ is a monomorphism we have $da = 0$ or $x = 0$. This proves that $\varphi^{(n+1)}_{m}$ is a monomorphism on the linear space spanned by free generators in $\mathcal{M}^{(n+1)}_{m}$ adjoined to $\mathcal{M}^{(n+1)}_{m-1}$. As by induction $\varphi^{(n+1)}_{m-1}$ is already a monomorphism on $\mathcal{M}^{(n+1)}_{m-1}$ and all the DGA's $\mathcal{M}^{(n+1)}_{m-1}$, $\mathcal{M}^{(n+1)}_{m}$, $\bar{\mathcal{M}}^{(n+1)}$ are free as GA's, it is seen that $\varphi^{(n+1)}_{m}$ is a monomorphism on $\mathcal{M}^{(n+1)}_{m}$. By induction φ is thus a monomorphism.

It remains to prove that φ is also an epimorphism. Suppose that $\bar{\mathcal{M}}^{(n+1)}_{m-1}$ $(m \geq 0)$ has already been proved to be in the image of φ and let us prove that this is also true for $\bar{\mathcal{M}}^{(n+1)}_{m}$. To see this, let us consider any element $\bar{x}$ of degree $n+1$ in $\bar{\mathcal{M}}^{(n+1)}_{m}$ but not in $\bar{\mathcal{M}}^{(n+1)}_{m-1}$. Then $d\bar{x} \in \bar{\mathcal{M}}^{(n+1)}_{m-1}$ so that, by induction hypothesis, there is some $y \in \mathcal{M}^{(n+1)}$ with $\varphi y = d\bar{x}$. As $\varphi dy = d\varphi y = dd\bar{x} = 0$ and φ is already proved to be a monomorphism, $dy = 0$ too or y is a cycle. As $\varphi y = d\bar{x} \sim 0$ in $\mathcal{M}$ and φ is an H-isomorphism $y \sim 0$ in $\mathcal{M}$ too, so $y = da$ for some a in $\mathcal{M}$. Now, $d(\bar{x} - \varphi a) = d\bar{x} - \varphi da = d\bar{x} - \varphi y = 0$ so $\bar{x} - \varphi a$ is a cycle. As φ is an H-isomorphism there is a cycle $x \in \mathcal{M}$ and an element $\bar{b} \in \bar{\mathcal{M}}$ such that $\varphi x = \bar{x} - \varphi a + d\bar{b}$. As $\deg \bar{b} = n$ and by induction hypothesis $\bar{\mathcal{M}}^{(n)} = \bar{\mathcal{M}}^{(n+1)}_{-1}$ is in the image of φ, there is some element $b \in \mathcal{M}^{(n)}$ such that $\varphi b = \bar{b}$. It follows that $\bar{x} = \varphi(x + a - db)$ for which $x + a - db \in \mathcal{M}^{(n+1)}$ being of degree $n+1$. It follows that $\bar{x}$ is in the image of φ. As the DGA $\bar{\mathcal{M}}^{(n+1)}_{m}$ is obtained from the DGA $\bar{\mathcal{M}}^{(n+1)}_{m-1}$ by adjoining free generators of degree $n+1$, it follows that $\bar{\mathcal{M}}^{(n+1)}_{m}$ is wholly in the image of φ. This completes the induction.

Next, we come to the following important

Lemma 2 (Lifting Lemma): Let

$$\varphi : \mathcal{B} \to \mathcal{A}$$

be an H-isomorphism of DGA's and $\mathcal{M}$ a minimal DGA . Then, to any DGA-morphism

$$\lambda : \mathcal{M} \to \mathcal{A}$$

there is some lifting DGA-morphism

$$\mu : \mathcal{M} \to \mathcal{B}$$

such that the following diagram is homotopically or $\simeq$-commutative:

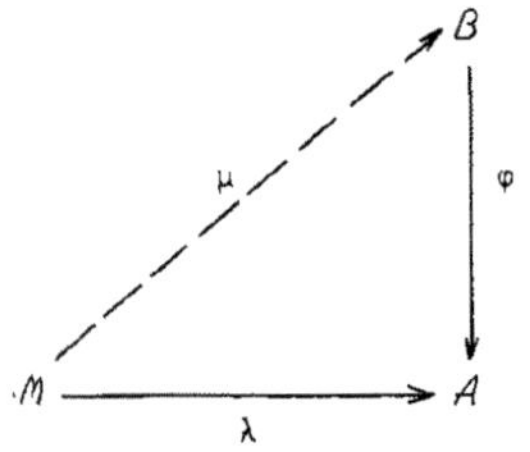

In other words,

$$\varphi\mu \simeq \lambda : \mathcal{M} \to \mathcal{A} .$$

Proof: We shall use the same notations $\mathcal{M}_m^{(n)}$ as before and define by induction a diagram $(D)_m^{(n)}$ verifying conditions $1_m^{(n)} - 4_m^{(n)}$ below:

$(D)_m^{(n)}$

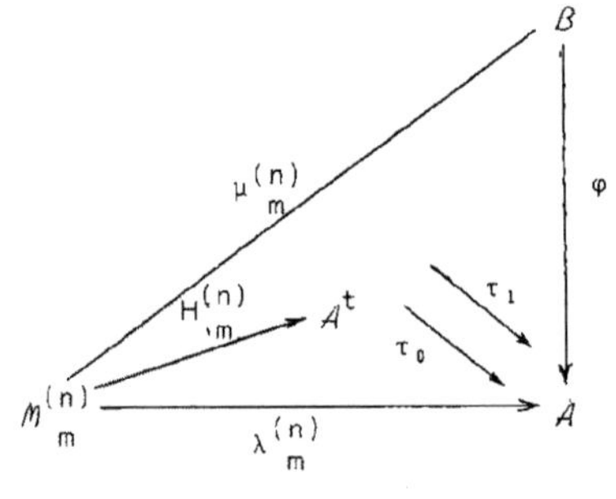

$1_m^{(n)}$. $\lambda_m^{(n)} = \lambda \mid \mathcal{M}_m^{(n)}$.

$2_m^{(n)}$. $H_m^{(n)}$, $\mu_m^{(n)}$ are DGA-morphisms .

$3_m^{(n)}$. $H_m^{(n)} \mid \mathcal{M}_{m-1}^{(n)} = H_{m-1}^{(n)}$,

$\mu_m^{(n)} \mid \mathcal{M}_{m-1}^{(n)} = \mu_{m-1}^{(n)}$.

$4^{(n)}_m$. The diagram $(D)^{(n)}_m$ is commutative or

$$\tau_1 H^{(n)}_m = \varphi\mu^{(n)}_m ,$$
$$\tau_0 H^{(n)}_m = \lambda^{(n)}_m .$$

Note that $\mathcal{M}^{(n)}_{-1}$ stands for $\mathcal{M}^{(n-1)}$ as before.

For $n = 0$ or $n = 1$, $m = 0$ the definition of $H^{(0)}$, $\mu^{(0)}$ is trivial. Suppose $(D)^{(n)}_{m-1}$ $(m \geq 0)$ has been defined with $1^{(n)}_{m-1} - 4^{(n)}_{m-1}$ verified. Define now $(D)^{(n)}_m$ as follows.

Let $\mathcal{M}^{(n)}_m = \mathcal{M}^{(n)}_{m-1} \otimes \text{Free}\ (z_\alpha)$ with $\deg z_\alpha = n$, $dz_\alpha \in \mathcal{M}^{(n)}_{m-1}$. For each $z = z_\alpha$ we have by Lemmas in §II 2,

$$\begin{aligned}
\varphi\mu^{(n)}_{m-1}\, dz &= \tau_1 H^{(n)}_{m-1}\, dz \\
&= (dI + Id + \tau_0)\, H^{(n)}_{m-1}\, dz \\
&= dIH^{(n)}_{m-1}\, dz + IdH^{(n)}_{m-1}\, dz + \lambda^{(n)}_{m-1}\, dz \\
&= d(IH^{(n)}_{m-1}\, dz + \lambda^{(n)}_m\, z) \\
&\sim 0 \text{ in } \mathcal{A} .
\end{aligned}$$

As φ is an H-isomorphism, we have

$$\mu^{(n)}_{m-1}\, dz \sim 0 \quad \text{in } \mathcal{B}$$

too, or for some $b \in \mathcal{B}$,

$$\mu^{(n)}_{m-1}\, dz = db .$$

It follows that

$$d\varphi b = \varphi db = \varphi\mu^{(n)}_{m-1}\, dz = d(IH^{(n)}_{m-1}\, dz + \lambda^{(n)}_m z)$$

so that

$$\varphi b - \lambda^{(n)}_m z - IH^{(n)}_{m-1}\, dz$$

is a cycle in $\mathcal{A}$. Since φ is an H-isomorphism there would exist a cycle $x \in \mathcal{B}$ and an element $a \in \mathcal{A}$ such that

$$dx = 0 \quad \text{and}$$

$$\varphi b - \lambda^{(n)}_m z - IH^{(n)}_{m-1}\, dz = \varphi x + da .$$

Define now

$$\mu^{(n)}_m z = b - x ,$$
$$H^{(n)}_m z = \lambda^{(n)}_m z + I_t H^{(n)}_{m-1} dz + d(ta) ,$$
$$\mu^{(n)}_m \mid \mathcal{M}^{(n)}_{m-1} = \mu^{(n)}_{m-1} ,$$
$$H^{(n)}_m \mid \mathcal{M}^{(n)}_{m-1} = H^{(n)}_{m-1} .$$

Then

$$d\mu^{(n)}_m z = db - dx = \mu^{(n)}_{m-1} dz = \mu^{(n)}_m dz$$

so that $\mu^{(n)}_m$ can be extended to a DGA-morphism

$$\mu^{(n)}_m : \mathcal{M}^{(n)}_m \to \mathcal{B} .$$

Similarly

$$\begin{aligned} dH^{(n)}_m z &= d\lambda^{(n)}_m z + dI_t H^{(n)}_{m-1} dz \\ &= \lambda^{(n)}_{m-1} dz + (1 - \tau_0 - I_t d) H^{(n)}_{m-1} dz \\ &= H^{(n)}_{m-1} dz = H^{(n)}_m dz , \end{aligned}$$

since $dH^{(n)}_{m-1} = H^{(n)}_{m-1} d$ and $\tau_0 H^{(n)}_{m-1} = \lambda^{(n)}_{m-1}$ by induction on $(D)^n_{m-1}$. It follows that $H^{(n)}_m$ can be extended to a DGA-morphism on $\mathcal{M}^{(n)}_m$ and $2^{(n)}_m$ is proved.

By using Lemmas in §II 2, condition $4^{(n)}_m$ can also easily be verified so that induction for $(D)^{(n)}_m$ is complete. Define now

$$\mu : \mathcal{M} \to \mathcal{A}$$

and

$$H : \mathcal{M} \to \mathcal{A}^t$$

by

$$\mu \mid \mathcal{M}^{(n)}_m = \mu^{(n)}_m ,$$
$$H \mid \mathcal{M}^{(n)}_m = H^{(n)}_m$$

for all n , m , then we have a commutative diagram below which proves the $\simeq$-commutativity $\varphi\mu \simeq \lambda$, as to be asserted:

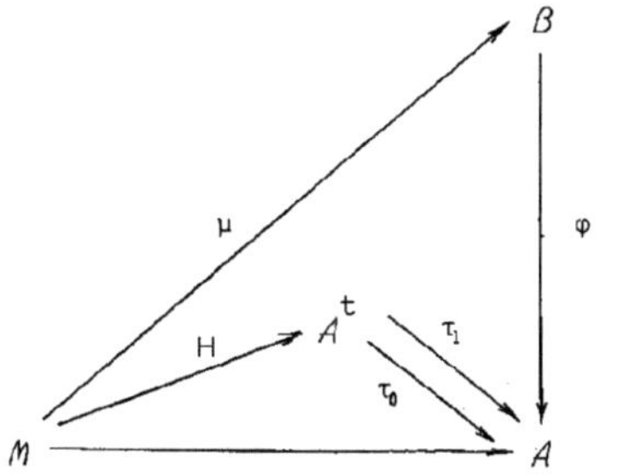

<u>Remark</u>: The DGA-morphism μ whose existence is asserted in the above Lemma 2 is unique only up to homotopy. For the proof we shall postpone to the next section.

<u>Theorem 1</u>: The minimal model $\mathcal{M}$ of a DGA $\mathcal{A}$ is unique up to isomorphism.

<u>Proof</u>: Let $\mathcal{M}, \overline{\mathcal{M}}$ be two minimal models of the DGA $\mathcal{A}$ with

$$\rho : \mathcal{M} \to \mathcal{A}$$

and

$$\rho : \overline{\mathcal{M}} \to \mathcal{A}$$

as associated minimal morphisms. By Lemma 2, there exists a DGA-morphism

$$\mu : \mathcal{M} \to \overline{\mathcal{M}}$$

such that

$$\rho \simeq \overline{\rho}\mu : \mathcal{M} \to \mathcal{A}$$

By Prop. 2 of §II 2, ρ and $\overline{\rho}\mu$ will induce identical H-morphisms

$$\rho_H = (\overline{\rho}\mu)_H = \overline{\rho}_H \mu_H : H(\mathcal{M}) \to H(\mathcal{A}) .$$

Since $\rho, \overline{\rho}$ are both H-isomorphisms or ρ_H, $\overline{\rho}_H$ are both isomorphisms, it follows that μ_H is also an isomorphism. By Lemma 1 μ is then itself an isomorphism

$$\mu : \mathcal{M} \approx \overline{\mathcal{M}} ,$$

as it will be proved.

<u>Theorem 2</u>: Let $\mathcal{M}$ be a minimal model of $\mathcal{A}$. Then any two associated minimal morphisms

$$\rho_0 , \rho_1 : \mathcal{M} \to \mathcal{A}$$

are identical up to homotopy in the sense that there exists a DGA-isomorphism i of $\mathcal{M}$ onto itself such that

$$\rho_1 i \simeq \rho_0 .$$

<u>Proof:</u> By the lifting lemma with $\mathcal{M}$ as $\mathcal{B}$, ρ_0, ρ_1 as λ and φ there exists a DGA-morphism $i : \mathcal{M} \to \mathcal{M}$ with $\rho_1 i \simeq \rho_0$. Since ρ_0, ρ_1 are both H-isomorphisms this is also true for i. By Lemma 1 i is then an isomorphism, as it will be proved.

The lifting lemma and the uniqueness theorem in this section furnish us a very useful method in determining the minimal model of a DGA by reducing the determination to some easier ones. First let us lay down the following

<u>Definition</u>: Two DGA's $\mathcal{A}$ and $\mathcal{B}$ will be said to be H-<u>equivalent</u> if there exists a series of DGA's $\mathcal{A}_0, \mathcal{A}_1, \ldots, \mathcal{A}_n$ and DGA-morphisms either $f_i : \mathcal{A}_i \to \mathcal{A}_{i+1}$ or $f_i : \mathcal{A}_{i+1} \to \mathcal{A}_i$ for $0 \le i \le n-1$ such that $\mathcal{A}_0 = \mathcal{A}$, $\mathcal{A}_n = \mathcal{B}$ and each f_i is an H-isomorphism.

Suppose now we are able to determine the minimal model $\mathcal{M}$ of the DGA $\mathcal{B} = \mathcal{A}_n$ with an associated minimal morphism $\rho_n : \mathcal{M} \to \mathcal{A}_n$. If f_{n-1} is in the direction $\mathcal{A}_n \to \mathcal{A}_{n-1}$ then $\rho_{n-1} = f_{n-1}\rho_n : \mathcal{M} \to \mathcal{A}_{n-1}$ clearly is an H-isomorphism so that $\mathcal{M}$ is also the minimal model of $\mathcal{A}_{n-1}$ with ρ_{n-1} an associated minimal morphism. If f_{n-1} is in the direction $\mathcal{A}_{n-1} \to \mathcal{A}_n$ then by the lifting lemma there will be some DGA-morphism $\rho_{n-1} : \mathcal{M} \to \mathcal{A}_{n-1}$ such that $f_{n-1}\rho_{n-1} \simeq \rho_n$. It follows by theorems in this section that f_{n-1} is again an H-isomorphism so that $\mathcal{M}$ is still the minimal model of $\mathcal{A}_{n-1}$ with ρ_{n-1} as some associated minimal morphism. This can be continued to finally getting an assoicated minimal morphism $\rho_0 : \mathcal{M} \to \mathcal{A}_0 = \mathcal{A}$ showing that $\mathcal{M}$ is also the minimal model of $\mathcal{A}$. The procedure thus reduces the determination of the minimal model of a DGA $\mathcal{A}$ to that of an H-equivalent one $\mathcal{B}$ which may be easier to do. This is a general <u>principle</u> frequently used and may be sketched as in the following diagram:

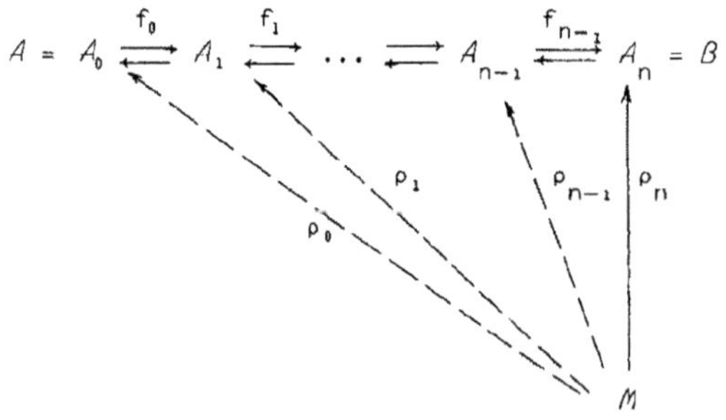

II.5 INDUCED MORPHISMS OF MINIMAL DGA's

In this section we shall prove the following

Theorem 1: For any DGA-morphism of DGA's

$$f : \mathcal{A} \to \mathcal{B}$$

with

$$\rho_{\mathcal{A}} : \min \mathcal{A} \to \mathcal{A} ,$$
$$\rho_{\mathcal{B}} : \min \mathcal{B} \to \mathcal{B}$$

as some associated minimal morphisms, there will be some induced DGA-morphism

$$\tilde{f} = \min f : \min \mathcal{A} \to \min \mathcal{B}$$

such that

$$f\rho_{\mathcal{A}} \simeq \rho_{\mathcal{B}}\tilde{f} : \min \mathcal{A} \to \mathcal{B} .$$

In this case we say that the diagram below is homotopically-commutative or $\simeq$-commutative:

$$(D) \qquad \begin{array}{ccc} \mathcal{A} & \xrightarrow{f} & \mathcal{B} \\ {\scriptstyle \rho_{\mathcal{A}}}\uparrow & & \uparrow{\scriptstyle \rho_{\mathcal{B}}} \\ \min \mathcal{A} & \xrightarrow{\tilde{f}} & \min \mathcal{B} \end{array}$$

Moreover, $\min f = \tilde{f}$ is unique up to homotopy in the sense that if $\tilde{f}' : \min \mathcal{A} \to \min \mathcal{B}$ is another such DGA-morphism verifying $f\rho_{\mathcal{A}} \simeq \rho_{\mathcal{B}}\tilde{f}'$, then $\tilde{f} \simeq \tilde{f}'$.

For the proof we shall first establish the following

Lemma 1 (Cancellation Lemma): Let $\mathcal{M},\mathcal{N}$ be minimal DGA's and

$$f_i : \mathcal{M} \to \mathcal{N} , \quad i = 0,1$$

be DGA-morphisms. Let $\mathcal{B}$ be a DGA with $\mathcal{N}$ as its minimal model and $g : \mathcal{N} \to \mathcal{B}$ an associated minimal morphism. If

$$gf_0 \simeq gf_1 : \mathcal{M} \to \mathcal{B} ,$$

then

$$f_0 \simeq f_1 : \mathcal{M} \to \mathcal{N} .$$

Proof: The hypothesis $gf_0 \simeq gf_1$ implies the existence of a DGA-morphism

$$H : \mathcal{M} \to \mathcal{B}^t$$

such that the following diagrams (D_i), $i = 0,1$, are commutative:

(D_i)

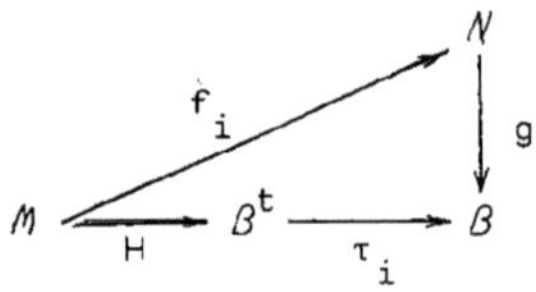

Introducing the trivial extension $\mathcal{N}^t$ of $\mathcal{N}$ and extending g to $g^t : \mathcal{N}^t \to \mathcal{B}^t$ as $g^t = g \otimes \text{ident.}$, we shall define a DGA-morphism $\tilde{H} : \mathcal{M} \to \mathcal{N}^t$ and complete the above commutative diagram to the following one

$(\tilde{D})$

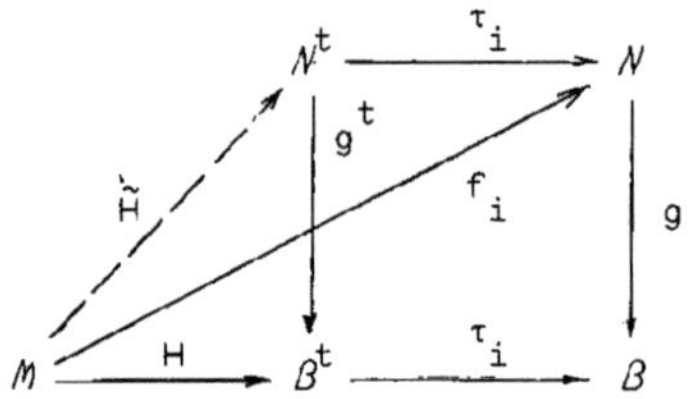

We shall do this by defining inductively the following commutative diagrams

$(D)^{(n)}_m$

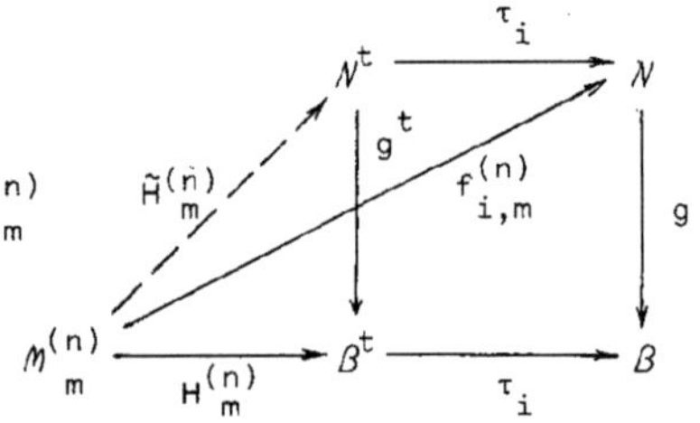

in which $H^{(n)}_m = H \mid \mathcal{M}^{(n)}_m$, $f^{(n)}_{i,m} = f_i \mid \mathcal{M}^{(n)}_m$, with meaning of $\mathcal{M}^{(n)}_m$ as before. For $n = 0$ or $n = 1$ and $m = 0$ the definition $\tilde{H}^{(n)}_m$ is trivial. Suppose now $(\tilde{D})^{(n)}_{m-1}$ has been defined and let us try to extend the definition to $(\tilde{D})^{(n)}_m$ with $\tilde{H}^{(n)}_m \mid \mathcal{M}^{(n)}_m = \tilde{H}^{(n)}_{m-1}$ as follows.

Let $\mathcal{M}^{(n)}_m = \mathcal{M}^{(n)}_{m-1} \otimes \text{Free}(z_\alpha)$ with $\deg z_\alpha = n$ and $dz_\alpha \in \mathcal{M}^{(n)}_{m-1}$. For each $z = z_\alpha$, $\tilde{H}^{(n)}_{m-1} dz$ is well-defined and by Lemmas in §II 2 we have

$$d(f^{(n)}_{1,m}z - f^{(n)}_{0,m}z - I\tilde{H}^{(n)}_{m-1}\,dz)$$
$$= (f^{(n)}_{1,m-1} - f^{(n)}_{0,m-1} - dI\tilde{H}^{(n)}_{m-1})\,dz$$
$$= (\tau_1 - \tau_0 - dI)\,\tilde{H}^{(n)}_{m-1}\,dz$$
$$= Id\tilde{H}^{(n)}_{m-1}\,dz$$
$$= 0\ .$$

Hence $f^{(n)}_{1,m}z - f^{(n)}_{0,m}z - I\tilde{H}^{(n)}_{m-1}\,dz$ is a cycle.

Let us remark that $Ig^t = gI$ from the very definition of I. Hence

$$g(f^{(n)}_{1,m}z - f^{(n)}_{0,m}z - I\tilde{H}^{(n)}_{m-1}\,dz)$$
$$= (\tau_1 - \tau_0)\,H^{(n)}_m z - gI\tilde{H}^{(n)}_{m-1}\,dz$$
$$= (dI + Id)\,H^{(n)}_m z - Ig^t\tilde{H}^{(n)}_{m-1}\,dz$$
$$= dIH^{(n)}_m z$$
$$\sim 0 \text{ in } \mathcal{B}\ .$$

Since g is an H-isomorphism

$$f^{(n)}_{1,m}z - f^{(n)}_{0,m}z - I\tilde{H}^{(n)}_{m-1}\,dz \sim 0 \text{ in } \mathcal{N}$$

too, so that for some $a \in \mathcal{N}$, we have

$$f^{(n)}_{1,m}z - f^{(n)}_{0,m}z - I\tilde{H}^{(n)}_{m-1}\,dz = da\ .$$

Define now

$$\tilde{H}^{(n)}_m \mid \mathcal{M}^{(n)}_{m-1} = \tilde{H}^{(n)}_{m-1},$$
$$\tilde{H}^{(n)}_m z = f^{(n)}_{0,m}z + I_t\tilde{H}^{(n)}_{m-1}\,dz + d(ta)\ .$$

It is easy to verify that

$$\tau_0\tilde{H}^{(n)}_m z = f_0 z\ ,$$
$$\tau_1\tilde{H}^{(n)}_m z = f_1 z\ ,$$

and

$$d\tilde{H}^{(n)}_m z = \tilde{H}^{(n)}_{m-1}\,dz = \tilde{H}^{(n)}_m\,dz\ .$$

It follows that $\tilde{H}^{(n)}_m$ can be extended to a DGA-morphism of $\mathcal{M}^{(n)}_m$ to $\mathcal{N}^t$ with diagram $(\tilde{D})^{(n)}_m$ commutative.

Defining now

$$\tilde{H} : \mathcal{M} \to \tilde{\mathcal{N}}$$

by $\tilde{H} \mid \mathcal{M}^{(n)}_m = \tilde{H}^{(n)}_m$ for all n,m we get then the commutative diagram $(\tilde{D})$ which shows that $f_0 \simeq f_1 : \mathcal{M} \to \mathcal{N}$ as to be proved.

As a corollary we have the following Lemma as a complement to the Lifting Lemma of §II 4.

Lemma 2: Let

$$\varphi : \mathcal{B} \to \mathcal{A}$$

be an H-isomorphism of DGA's and $\mathcal{M}$ a minimal DGA. Then, to any DGA-morphism

$$\lambda : \mathcal{M} \to \mathcal{A}$$

there is some lifting DGA-morphism

$$\mu : \mathcal{M} \to \mathcal{B}$$

such that

$$\varphi\mu \simeq \lambda : \mathcal{M} \to \mathcal{A}$$

and any two such DGA-morphisms μ, μ' are homotopic.

Now we come to the

Proof of Theorem 1: By the Lifting Lemma of §II 4 DGA-morphism

$$\min f : \min \mathcal{A} \to \min \mathcal{B}$$

exists with

$$\rho_{\mathcal{B}} \cdot \min f \simeq f \rho_{\mathcal{A}} .$$

By the Cancellation Lemma 1 and Lemma 2 above $\min f$ is then unique up to homotopy as to be proved.

Theorem 1 asserts the existence of induced morphism $\min f : \min \mathcal{A} \to \min \mathcal{B}$ for a given DGA-morphism $f : \mathcal{A} \to \mathcal{B}$ which renders the corresponding diagram only homotopically commutative but not commutative in the strict sense which is in general impossible. As a simple example let us consider the following:

Ex. $\mathcal{A}$ = Polym(a) , deg a = n even , da = 0 ,

$\mathcal{B}$ = Extr(b) $\otimes$ Polym(c)

as graded algebra, with

deg b = n − 1 , deg c = n ,

db = c , dc = 0 ,

and

$f : \mathcal{A} \to \mathcal{B}$

given by

f(a) = c .

Then

$\min \mathcal{A} = \mathcal{A}$,

$\min \mathcal{B} = \underline{k}$,

so no DGA-morphism

$\tilde{f} : \min \mathcal{A} \to \min \mathcal{B}$

can exist for which the diagram can be made commutative.

However, there is an important case for which min f exists with corresponding diagram strictly commutative, viz. the case for which $f : \mathcal{A} \to \mathcal{B}$ is onto. In fact, we have the following

Theorem 2: If the DGA-morphism

$f : \mathcal{A} \to \mathcal{B}$

is an epimorphism, then for any associated minimal morphism

$\rho_{\mathcal{A}} : \min \mathcal{B} \to \mathcal{B}$

there is a minimal morphism

$\rho_{\mathcal{A}} : \min \mathcal{A} \to \mathcal{A}$

of $\mathcal{A}$ and an induced DGA-morphism

$$\tilde{f} = \min f \; : \; \min\mathcal{A} \to \min\mathcal{B}$$

such that the diagram (D) is strictly commutative.

Proof: To simplify the notation we shall write $\mathcal{M}$ and $\mathcal{N}$ for $\min\mathcal{A}$ and $\min\mathcal{B}$. Now $\mathcal{N}$ and $\rho_{\mathcal{B}}$ being given, we shall construct the minimal model $\mathcal{M}$ as well as $\rho_{\mathcal{A}}$ and $\tilde{f} = \min f$ by closely following the steps in the proof of the existence theorem in §II 3, with due modifications, in order to make the diagram (D) commutative.

Thus, we shall successively construct a sequence of DGA's.

$$\underline{k} = \mathcal{M}^{(0)} \subset \mathcal{M}^{(1)} \subset \ldots \subset \mathcal{M}^{(n)} \subset \ldots$$

as well as extending sequence of DGA-morphisms

$$\rho_{\mathcal{A}}^{(n)} : \mathcal{M}^{(n)} \to \mathcal{A} ,$$
$$\tilde{f}^{(n)} : \mathcal{M}^{(n)} \to \mathcal{N}$$

to make the diagram $(\bar{D})^{(n)}$ below strictly commutative.

$$(\bar{D})^{(n)} \qquad \begin{array}{ccccc} \mathcal{A} & \xrightarrow{f} & \mathcal{B} & \longrightarrow & 0 \\ \uparrow{\scriptstyle \rho_{\mathcal{A}}^{(n)}} & & \uparrow{\scriptstyle \rho_{\mathcal{B}}} & & \\ \mathcal{M}^{(n)} & \xrightarrow{\tilde{f}^{(n)}} & \mathcal{N} & & \end{array}$$

To begin with, we shall define $\tilde{f}^{(0)} \mid \mathcal{M}^{(0)}$ and $\rho_{\mathcal{A}}^{(0)} \mid \mathcal{M}^{(0)}$ to be simply the identity since $\mathcal{M}^{(0)} = \underline{k}$.

Suppose now $\mathcal{M}^{(n)}$ and $\rho_{\mathcal{A}}^{(n)}$. $\tilde{f}^{(n)}$ have already been defined with corresponding diagram $(\bar{D})^{(n)}$ commutative and let us try to extend to a diagram $(\bar{D})^{(n+1)}$ as follows.

Let us first consider the commutative diagram of morphisms below:

$$\begin{array}{ccccccccc} 0 & \longrightarrow & H_{n+1}(\mathcal{M}^{(n)}) & \xrightarrow{\rho^{(n)}_{\mathcal{A},H_{n+1}}} & H_{n+1}(\mathcal{A}) & \longrightarrow & \operatorname{Coker} \rho^{(n)}_{\mathcal{A},H_{n+1}} & \longrightarrow & 0 \\ & & \downarrow{\scriptstyle \tilde{f}^{(n)}_{H_{n+1}}} & & \downarrow{\scriptstyle f_{H_{n+1}}} & & \downarrow & & \\ 0 & \longrightarrow & H_{n+1}(\mathcal{N}) & \xrightarrow{\rho^{(n)}_{\mathcal{B},H_{n+1}}} & H_{n+1}(\mathcal{B}) & \longrightarrow & \operatorname{Coker} \rho^{(n)}_{\mathcal{B},H_{n+1}} & \longrightarrow & 0 \end{array}$$

If $\operatorname{Coker} \rho^{(n)}_{\mathcal{A},H_{n+1}} = 0$, then we put $\mathcal{M}_0^{(n+1)} = \mathcal{M}^{(n)}$ and define

$$\rho_{A,0}^{(n+1)} : M_0^{(n+1)} \to A$$

$$\tilde{f}_0^{(n+1)} : M_0^{(n+1)} \to N$$

simply as $\rho^{(n)}$ and $\tilde{f}^{(n)}$. Otherwise take a $\underline{k}$-basis $\{X_j^{(n+1)} \bmod \operatorname{Im} \rho_{A,H_{n+1}}^{(n)}\}$ of Coker $\rho_{A,H_{n+1}}^{(n)}$ and a cycle $x_j^{(n+1)} \in A$ in each class $X_j^{(n+1)}$.
Since $f x_j^{(n+1)}$ is a cycle of B and ρ_B is an H-isomorphism there should exist some cycle $y_j^{(n+1)}$ of N and some element $b_j^{(n+1)}$ of B such that

$$f x_j^{(n+1)} = \rho_B y_j^{(n+1)} + d b_j^{(n+1)} .$$

Since f is <u>onto</u> there is some $a_j^{(n+1)}$ in A with

$$b_j^{(n+1)} = f a_j^{(n+1)}$$

so that

$$f(x_j^{(n+1)} - d a_j^{(n+1)}) = \rho_B y_j^{(n+1)} .$$

Now, introduce new generators $\xi_j^{(n+1)}$ to $M^{(n)}$ and set

$$M_0^{(n+1)} = M^{(n)} \otimes \text{Free}\,(\xi_j^{(n+1)})$$

with

$$\deg \xi_j^{(n+1)} = n+1 , \quad d\xi_j^{(n+1)} = 0 .$$

We also define DGA-morphisms

$$\rho_{A,0}^{(n+1)} : M_0^{(n+1)} \to A ,$$

$$\tilde{f}_0^{(n+1)} : M_0^{(n+1)} \to N$$

by setting

$$\rho_{A,0}^{(n+1)} \mid M^{(n)} = \rho^{(n)} ,$$

$$\tilde{f}_0^{(n+1)} \mid M^{(n)} = \tilde{f}^{(n)} ,$$

$$\rho_{A,0}^{(n+1)}(\xi_j^{(n+1)}) = x_j^{(n+1)} - d a_j^{(n+1)} ,$$

$$\tilde{f}_0^{(n+1)}(\xi_j^{(n+1)}) = y_j^{(n+1)} .$$

Then, the diagram $(\bar{D})_0^{(n+1)}$ is strictly commutative:

$$(\bar{D})_0^{(n+1)} \qquad \begin{array}{ccccc} \mathcal{A} & \xrightarrow{f} & \mathcal{B} & \longrightarrow & 0 \\ {\scriptstyle \rho_{\mathcal{A},0}^{(n+1)}}\uparrow & & \uparrow{\scriptstyle \rho_{\mathcal{B}}} & & \\ \mathcal{M}_0^{(n+1)} & \xrightarrow{f_0^{(n+1)}} & \mathcal{N} & & \end{array}$$

We now extend $\mathcal{M}_0^{(n+1)}$ to an increasing sequence of DGA's

$$\mathcal{M}_0^{(n+1)} \subset \mathcal{M}_1^{(n+1)} \subset \dots \subset \mathcal{M}_m^{(n+1)} \subset \dots$$

and also extending sequences of DGA-morphisms

$$\rho_{\mathcal{A},m}^{(n+1)} : \mathcal{M}_m^{(n+1)} \to \mathcal{A},$$

$$\tilde{f}_m^{(n+1)} : \mathcal{M}_m^{(n+1)} \to \mathcal{N}$$

to make the diagram $(\bar{D})_m^{(n+1)}$ below strictly commutative, as follows:

$$(\bar{D})_m^{(n+1)} \qquad \begin{array}{ccccc} \mathcal{A} & \xrightarrow{f} & \mathcal{B} & \longrightarrow & 0 \\ {\scriptstyle \rho_{\mathcal{A},m}^{(n+1)}}\uparrow & & \uparrow{\scriptstyle \rho_{\mathcal{B}}} & & \\ \mathcal{M}_m^{(n+1)} & \xrightarrow{\tilde{f}_m^{(n+1)}} & \mathcal{N} & & \end{array}$$

Suppose that $\mathcal{M}_m^{(n+1)}$, $\rho_{\mathcal{A},m}^{(n+1)}$ and $\tilde{f}_m^{(n+1)}$ have already been defined to meet the requirements and consider the commutative diagram below

$$\begin{array}{ccccccc} 0 & \longrightarrow & \operatorname{Ker} \rho_{\mathcal{A},m,H_{n+2}}^{(n+1)} & \longrightarrow & H_{n+2}(\mathcal{M}_m^{(n+1)}) & \xrightarrow{\rho_{\mathcal{A},m,H_{n+2}}^{(n+1)}} & H_{n+2}(\mathcal{A}) \\ & & \downarrow & & \downarrow{\scriptstyle \tilde{f}_{m,H_{n+2}}^{(n+1)}} & & \downarrow{\scriptstyle f_{H_{n+2}}} \\ 0 & \longrightarrow & \operatorname{Ker} \rho_{\mathcal{B},H_{n+2}}^{(n+1)} & \longrightarrow & H_{n+2}(\mathcal{N}) & \xrightarrow{\rho_{\mathcal{B},H_{n+2}}^{(n+1)}} & H_{n+2}(\mathcal{B}) \end{array}$$

If $\operatorname{Ker} \rho_{\mathcal{A},m,H_{n+2}}^{(n+1)} = 0$ then the construction will be stopped. Otherwise let $\{\Gamma_{mj}^{(n+1)}\}$ be a $\underline{k}$-basis of $\operatorname{Ker} \rho_{\mathcal{A},m,H_{n+2}}^{(n+1)}$ and take a cycle $\gamma_{mj}^{(n+1)} \in \mathcal{M}_m^{(n+1)}$ in each $\Gamma_{mj}^{(n+1)}$.

Then

$$\rho_{\mathcal{A},m}^{(n+1)} \gamma_{mj}^{(n+1)} = dc_{mj}^{(n+1)} \sim 0 \quad \text{in } \mathcal{A}$$

for some $c_{mj}^{(n+1)}$ of $\mathcal{A}$.

Now

$$\rho_{\mathcal{B}}\tilde{f}_m^{(n+1)}\,\gamma_{mj}^{(n+1)} = f\,\rho_{\mathcal{A},m}^{(n+1)}\,\gamma_{mj}^{(n+1)}$$

$$= f\,dc_{mj}^{(n+1)} = dfc_{mj}^{(n+1)}$$

$$\sim 0 \quad \text{in } \mathcal{B}\ .$$

Since $\rho_{\mathcal{B}}$ is an H-isomorphism we have

$$\tilde{f}_m^{(n-1)}\,\gamma_{mj}^{(n+1)} = dy_{mj}^{(n+1)} \sim 0 \quad \text{in } \mathcal{N}$$

for some $y_{mj}^{(n+1)} \in \mathcal{N}$. It follows that

$$d(\rho_{\mathcal{B}}y_{mj}^{(n+1)} - f\,c_{mj}^{(n+1)}) = 0$$

or $\rho_{\mathcal{B}}y_{mj}^{(n+1)} - fc_{mj}^{(n+1)}$ is a cycle in $\mathcal{B}$.

Again, since $\rho_{\mathcal{B}}$ is an H-isomorphism we should have

$$\rho_{\mathcal{B}}y_{mj}^{(n+1)} - f\,c_{mj}^{(n+1)} = \rho_{\mathcal{B}}u_{mj}^{(n+1)} + db_{mj}^{(n+1)}$$

for some cycle $u_{mj}^{(n+1)}$ in $\mathcal{N}$ and some element $b_{mj}^{(n+1)}$ in $\mathcal{B}$.

Since f is <u>onto</u> there should exist some element $a_{mj}^{(n+1)} \in \mathcal{A}$ such that

$$f\,a_{mj}^{(n+1)} = b_{mj}^{(n+1)}\ .$$

Now, introduce generators $\zeta_{mj}^{(n+1)}$ to $\mathcal{M}_m^{(n+1)}$ to form DGA

$$\mathcal{M}_{m+1}^{(n+1)} = \mathcal{M}_m^{(n+1)} \otimes \text{Free}\,(\zeta_{mj}^{(n+1)})$$

with gradation and differential given by

$$\deg \zeta_{mj}^{(n+1)} = n + 1\ ,$$

$$d\zeta_{mj}^{(n+1)} = \gamma_{mj}^{(n+1)}\ .$$

Now, extend morphisms $\rho_{\mathcal{A},m}^{(n+1)}$, $\tilde{f}_m^{(n+1)}$ to

$$\rho_{\mathcal{A},m+1}^{(n+1)} : \mathcal{M}_{m+1}^{(n+1)} \to \mathcal{A}\ ,$$

$$\tilde{f}_{m+1}^{(n+1)} : \mathcal{M}_{m+1}^{(n+1)} \to \mathcal{N}$$

by setting

$$\rho^{(n+1)}_{A,m+1} \mid \mathcal{M}^{(n+1)}_{m} = \rho^{(n+1)}_{A,m} ,$$

$$\tilde{f}^{(n+1)}_{m+1} \mid \mathcal{M}^{(n+1)}_{m} = \tilde{f}^{(n+1)}_{m} ,$$

$$\rho^{(n+1)}_{A,m+1}(\zeta^{(n+1)}_{mj}) = c^{(n+1)}_{mj} + da^{(n+1)}_{mj} ,$$

$$\tilde{f}^{(n+1)}_{m+1}(\zeta^{(n+1)}_{mj}) = y^{(n+1)}_{mj} - u^{(n+1)}_{mj} .$$

Then, it is easy to see that $\rho^{(n+1)}_{A,m+1}$ and $f^{(n+1)}_{m+1}$ commutes with d so that they can be extended to DGA-morphisms on $\mathcal{M}^{(n+1)}_{m+1}$. In this way, we get the commutative diagram $(\bar{D})^{(n+1)}_{m+1}$ as required.

Proceeding now in the same manner indefinitely, if necessary, to get a finite or infinite sequence of DGA's

$$\mathcal{M}^{(n+1)}_{0} \subset \mathcal{M}^{(n+1)}_{1} \subset \dots \subset \mathcal{M}^{(n+1)}_{m} \subset \dots$$

as well as sequences of DGA-morphisms $\rho^{(n+1)}_{A,m}$, $\tilde{f}^{(n+1)}_{m}$.

Now, define $\mathcal{M}^{(n+1)}$ as the union of all $\mathcal{M}^{(n+1)}_{m}$ and then $\mathcal{M}$ as the union of all $\mathcal{M}^{(n)}$ with DGA-morphisms

$$\rho_{A} : \mathcal{M} \to A ,$$

$$\tilde{f} : \mathcal{M} \to \mathcal{N}$$

defined by their restrictions on respective sub-DGA's $\mathcal{M}^{(n)}_{m}$, then we get the commutative diagram (D) as required. This proves the theorem.

II.6 SOME AUXILIARY THEOREMS ABOUT TWISTED PRODUCTS

In what follows, A,B are DGA's with differentials d_A and d_B. For the usual tensor product $A \otimes B$ the differential, denoted by $d_{\otimes}$, is given by (a,b homogeneous $\in A,B$ respectively).

(1) $$d_{\otimes}(a \otimes b) = d_A a \otimes b + (-1)^{\deg a} \cdot a \otimes d_B b .$$

Suppose that in $A \otimes B$ there is another differential

(2) $$d = d_{\otimes} + d_{\tau}$$

such that

(3) $\quad d_\tau a = 0$,

(4) $\quad d_\tau b \in \mathcal{A}^+ \otimes \mathcal{B}$.

It is to be understood that d_τ should verify some conditions in order to make d in (2) really a differential.

<u>Definition</u>: The tensor product $\mathcal{A} \otimes \mathcal{B}$ of the DGA's $\mathcal{A},\mathcal{B}$ with a differential d given by (2)-(4) is called a <u>twisted product</u> of $\mathcal{A},\mathcal{B}$ and will sometimes be denoted as $\mathcal{A} \otimes_\tau \mathcal{B}$ to make evidence the difference d_τ between the differential d from the usual one $d_\otimes$. The differential d in $\mathcal{A} \otimes_\tau \mathcal{B}$ is then said to be a <u>twisted differential</u> with d_τ as its <u>twisted part</u>.

<u>Definition</u>: The morphism

$$\varepsilon_\tau : \mathcal{A} \otimes_\tau \mathcal{B} \to \mathcal{B}$$

defined by sending $a \otimes b$ to 0 or ab accordingly as deg a is > 0 or = 0, clearly is a DGA-morphism and will be called the <u>restriction</u> of the twisted product $\mathcal{A} \otimes_\tau \mathcal{B}$.

<u>Definition</u>: Let $\mathcal{A}' \otimes_{\tau'} \mathcal{B}'$ be another twisted product of DGA's $\mathcal{A}', \mathcal{B}'$ with $d_{\tau'}$ as the twisted part of the twisted differential. Then, a DGA-morphism

$$\tilde{\rho} : \mathcal{A}' \otimes_{\tau'} \mathcal{B}' \to \mathcal{A} \otimes_\tau \mathcal{B}$$

will be said to be one <u>respecting</u> the twisted differentials if $\tilde{\rho}(\mathcal{A}') \subset \mathcal{A}$, and there is a commutative diagram of DGA-morphisms below

$$\begin{array}{ccccc}
\mathcal{A} & \xrightarrow{\iota_{\mathcal{A}}} & \mathcal{A} \otimes_\tau \mathcal{B} & \xrightarrow{\varepsilon_\tau} & \mathcal{B} \\
\uparrow{\scriptstyle \omega} & & \uparrow{\scriptstyle \tilde{\rho}} & & \uparrow{\scriptstyle \rho} \\
\mathcal{A}' & \xrightarrow{\iota_{\mathcal{A}'}} & \mathcal{A}' \otimes_{\tau'} \mathcal{B}' & \xrightarrow{\varepsilon_{\tau'}} & \mathcal{B}'
\end{array}$$

In the diagram both right horizontal arrows are restrictions and both left ones are inclusions. The DGA-morphisms ρ and ω will then be called the <u>restriction</u> and <u>induction</u> of $\tilde{\rho}$ respectively.

<u>Remark</u>: These terminologies are adapted to fit the study of fibrations in which $\mathcal{A},\mathcal{B}$

and $\mathcal{A} \otimes_\tau \mathcal{B}$ are DGA's connected with the base, the fiber and the fiber-space of a fibration. Similarly, for $\mathcal{A}', \mathcal{B}'$ and $\mathcal{A}' \otimes_\tau \mathcal{B}'$. The DGA-morphism w is then connected with one induced from certain map of the base spaces.

Let $\mathcal{M}, \mathcal{N}$ now be minimal models of $\mathcal{A}, \mathcal{B}$ with some minimal morphisms

$$\rho_{\mathcal{A}} : \mathcal{M} \to \mathcal{A}, \qquad \rho_{\mathcal{B}} : \mathcal{N} \to \mathcal{B}.$$

We are interested in the existence of some twisted product $\mathcal{M} \otimes_\sigma \mathcal{N}$ such that there is some DGA-morphism $\mathcal{M} \otimes_\sigma \mathcal{N} \to \mathcal{A} \otimes_\tau \mathcal{B}$ inducing an H-isomorphism and respecting the twisted differentials with $\rho_{\mathcal{A}}, \rho_{\mathcal{B}}$ as the respective induction and restriction.

Proposition 1: Let

$$\mathcal{C} = \mathcal{A} \otimes_\tau \mathcal{N},$$

$$\mathcal{P} = \mathcal{M} \otimes_\sigma \mathcal{N}$$

be twisted products with twisted parts of the twisted differentials d_τ and d_σ respectively. If some DGA-morphism

$$\tilde{\rho} : \mathcal{P} \to \mathcal{C}$$

respects the twisted differentials with the minimal morphism $\rho_{\mathcal{A}} : \mathcal{M} \to \mathcal{A}$ as the induction and ident. : $\mathcal{N} \to \mathcal{N}$ as the restriction, then $\tilde{\rho}$ is an H-isomorphism.

Proof: Let the minimal DGA $\mathcal{N}$ be the union of sub-DGA's $\mathcal{N}^{(n)}$, $n \geq 0$, and each $\mathcal{N}^{(n)}$ be the union of sub-DGA's $\mathcal{N}^{(n)}_m$'s, $m \geq -1$, with $\mathcal{N}^{(n)}_{-1} = \mathcal{N}^{(n-1)}$ as in the preceding sections. Set $(n \geq 0,\ m \geq -1)$

$$\mathcal{C}^{(n)} = \mathcal{A} \otimes_\tau \mathcal{N}^{(n)}, \qquad \mathcal{C}^{(n)}_m = \mathcal{A} \otimes_\tau \mathcal{N}^{(n)}_m,$$

$$\mathcal{P}^{(n)} = \mathcal{M} \otimes_\sigma \mathcal{N}^{(n)}, \qquad \mathcal{P}^{(n)}_m = \mathcal{M} \otimes_\sigma \mathcal{N}^{(n)}_m,$$

$$\tilde{\rho}^{(n)} = \tilde{\rho} \mid \mathcal{P}^{(n)} : \quad \mathcal{P}^{(n)} \to \mathcal{C}^{(n)}$$

$$\tilde{\rho}^{(n)}_m = \tilde{\rho} \mid \mathcal{P}^{(n)}_m : \quad \mathcal{P}^{(n)}_m \to \mathcal{C}^{(n)}_m .$$

We shall show by induction that for all $n \geq 0$, $m \geq -1$, $\tilde{\rho}^{(n)}_m$ and $\tilde{\rho}^{(n)}$ are H-isomorphisms.

Since $\tilde{\rho}^{(0)}$ is the same as $\rho_{\mathcal{A}} : \mathcal{M} \to \mathcal{A}$, it clearly is an H-isomorphism. Therefore, suppose for $n \geq 1$, $m \geq -1$, $\tilde{\rho}^{(n)}_m$ is an H-isomorphism and let us proceed to prove that $\tilde{\rho}^{(n)}_{m+1}$ is also an H-isomorphism as follows.

Let $\mathcal{N}^{(n)}_{m+1} = \mathcal{N}^{(n)}_{m} \otimes \mathrm{Free}\,(y_\alpha)$ with $d_{\mathcal{N}} y_\alpha \in \mathcal{N}^{(n)}_{m}$. Then $\mathcal{C}^{(n)}_{m+1} = \mathcal{C}^{(n)}_{m} \otimes \mathrm{Free}\,(y_\alpha)$ with $d_{\mathcal{C}} y_\alpha \in \mathcal{C}^{(n)}_{m}$. Consider any term of the form $cY \in \mathcal{C}^{(n)}_{m+1}$ in which $c \in \mathcal{C}^{(n)}_{m}$, while Y is a product of powers in the y'_αs. The degree of Y will then be denoted by $\deg_y(cY)$. All linear combinations of terms of same $\deg_y = k$ will then form a sub-graded-module of $\mathcal{C}^{(n)}_{m+1}$ which will become a differential graded module $\mathcal{C}'_k$ by introducing a differential d' given by $d'(cY) = d_{\mathcal{C}}c \cdot Y$, for cY above. Now, consider the sub-graded-module of $\mathcal{C}^{(n)}_{m+1}$ consisting of elements of the form

$$z = \sum_{j=0}^{k} z_j ,$$

in which each z_j is a linear combination of terms of $\deg_y = j$. It is clear that $d_{\mathcal{C}}$ will turn it into a differential graded module to be denoted by, say $\tilde{\mathcal{C}}_k$.

It is easy to see that the morphism $z \to z_k$ respects the differentials, and thus we have an exact sequence of differential graded modules

$$0 \longrightarrow \tilde{\mathcal{C}}_{k-1} \longrightarrow \tilde{\mathcal{C}}_k \longrightarrow \mathcal{C}'_k \longrightarrow 0 .$$

Similarly, let us form analogous differential graded modules in using $\mathcal{P}^{(n)}_m$ instead of $\mathcal{C}^{(n)}_m$, etc. Then we shall get analogous exact sequences in the $\mathcal{P}$'s, connected with the above one by natural morphisms, so that we have a commutative diagram below:

$$\begin{array}{ccccccccc}
0 & \longrightarrow & \tilde{\mathcal{C}}_{k-1} & \longrightarrow & \tilde{\mathcal{C}}_k & \longrightarrow & \mathcal{C}'_k & \longrightarrow & 0 \\
 & & \uparrow{\scriptstyle \tilde{\rho}_{k-1}} & & \uparrow{\scriptstyle \tilde{\rho}_k} & & \uparrow{\scriptstyle \rho'} & & \\
0 & \longrightarrow & \tilde{\mathcal{P}}_{k-1} & \longrightarrow & \tilde{\mathcal{P}}_k & \longrightarrow & \mathcal{P}'_k & \longrightarrow & 0
\end{array}$$

In the diagram ρ', for example, is defined by $\rho'(\gamma Y) = \tilde{\rho}^{(n)}_m \gamma \cdot Y$, where $\gamma \in \mathcal{P}^{(n)}_m$ and Y is a product of powers in y'_αs with degree $= k$. Remark that the commutativity of the right-hand square follows from the condition that the restrictions of $\tilde{\rho}$ is the identity on $\mathcal{N}$.

Now, $\tilde{\mathcal{C}}_0 = \mathcal{C}^{(n)}_m$, $\tilde{\mathcal{P}}_0 = \mathcal{P}^{(n)}_m$ as differential graded modules, so that by induction hypothesis $\tilde{\rho}_{0,H} : H_{\oplus}(\tilde{\mathcal{P}}_0) \approx H_{\oplus}(\tilde{\mathcal{C}}_0)$. For the same reason it is clear that $\rho'_H : H_{\oplus}(\mathcal{P}'_k) \approx H_{\oplus}(\mathcal{C}'_k)$. By usual arguments in using five-lemma on the homology sequences, arising from the above commutative diagrams, we then deduce $\tilde{\rho}_k : H_{\oplus}(\tilde{\mathcal{P}}_k) \approx H_{\oplus}(\tilde{\mathcal{C}}_k)$ for all k. This implies clearly, that

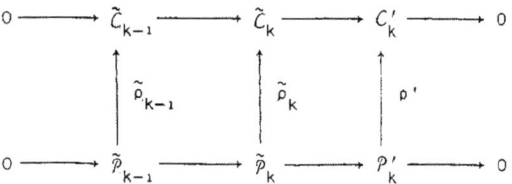

$$\tilde{\rho}^{(n)}_{m+1,H} : H_{\oplus}(\tilde{\mathcal{P}}^{(n)}_{m+1}) \approx H_{\oplus}(\tilde{\mathcal{C}}^{(n)}_{m+1}) .$$

Since the morphism $\tilde{\rho}^{(n)}_{m+1}$ is multiplicative, we therefore have also multiplicative isomorphism

$$\tilde{\rho}^{(n)}_{m+1,H} : H(\tilde{\mathcal{P}}^{(n)}_{m+1}) \approx H(\tilde{\mathcal{C}}^{(n)}_{m+1}) ,$$

or $\tilde{\rho}^{(n)}_{m+1,H}$ is an H-isomorphism.

Since each $\tilde{\mathcal{P}}^{(n)}$ (resp. $\tilde{\mathcal{C}}^{(n)}$) is the union of all $\tilde{\mathcal{P}}^{(n)}_m$ (resp. $\tilde{\mathcal{C}}^{(n)}_m$) and $\tilde{\mathcal{P}}$ (resp. $\tilde{\mathcal{C}}$) is the union of all $\tilde{\mathcal{P}}^{(n)}$ (resp. $\tilde{\mathcal{C}}^{(n)}$), we finally get by induction, that $\tilde{\rho}$ is an H-isomorphism, as to be proved.

Proposition 2: Let the twisted product $\mathcal{C} = \mathcal{A} \otimes_\tau \mathcal{N}$ and minimal morphism $\rho_{\mathcal{A}} : \mathcal{M} \to \mathcal{A}$ be as before. Then we can introduce in the tensor product $\mathcal{M} \otimes \mathcal{N}$ some twisted differential with twisted part d_σ, such that for the resulting twisted product $\mathcal{M} \otimes_\sigma \mathcal{N}$ there is some DGA-morphism

$$\tilde{\rho} : \mathcal{M} \otimes_\sigma \mathcal{N} \to \mathcal{A} \otimes_\tau \mathcal{N}$$

which respects the twisted differentials and has $\rho_{\mathcal{A}} : \mathcal{M} \to \mathcal{A}$ as the induction, and ident. : $\mathcal{N} \to \mathcal{N}$ as the restriction.

Proof: Let us use the same notations as in the proof of the preceding proposition. We shall proceed by induction in introducing twisted differential into $\mathcal{M} \otimes_\sigma \mathcal{N}^{(n)}_m$ and $\mathcal{M} \otimes_\sigma \mathcal{N}^{(n)}$ successively, beginning from the trivial case $n = 0$. Therefore, suppose $\mathcal{P}^{(n)}_m = \mathcal{M} \otimes_\sigma \mathcal{N}^{(n)}_m$ and $\tilde{\rho}^{(n)}_m : \mathcal{P}^{(n)}_m \to \mathcal{C}^{(n)}_m$, respecting the twisted differentials have already been defined, and proceed to define $d_\sigma y_\alpha$ and $\tilde{\rho}^{(n)}_{m+1}$ as follows.

Remark that $d_{\mathcal{C}} y_\alpha$ is a cycle in $\mathcal{C}^{(n)}_m$. By Prop. 1, $\tilde{\rho}^{(n)}_m$ is an H-isomorphism. Hence, we have

$$d_{\mathcal{C}}\, y_\alpha = \tilde{\rho}^{(n)}_m \zeta_\alpha + d_{\mathcal{C}} c_\alpha$$

for some cycle ζ_α in $\mathcal{P}^{(n)}_m$ and some element $c_\alpha \in \mathcal{C}^{(n)}_m$. Extend now $\tilde{\rho}^{(n)}_m \mid \mathcal{P}^{(n)}_m$ to $\tilde{\rho}^{(n)}_{m+1} : \mathcal{P}^{(n)}_{m+1} \to \mathcal{C}^{(n)}_{m+1}$ and extend $d_{\mathcal{P}}$ in $\mathcal{P}^{(n)}_{m+1}$ by setting

$$\tilde{\rho}^{(n)}_{m+1} y_\alpha = y_\alpha - c_\alpha ,$$

and

$$d_{\mathcal{P}}\, y_\alpha = \zeta_\alpha , \qquad d_{\mathcal{P}}\, y_\alpha = d_{\otimes} y_\alpha + d_\sigma\, y_\alpha .$$

Then it is easy to verify that $d_{\mathcal{P}}$ is a twisted differential in $\mathcal{P}^{(n)}_{m+1}$ and $\tilde{\rho}^{(n)}_{m+1}$ respects the twisted differentials. It is also clear that the restriction of

$\tilde{\rho}^{(n)}_{m+1}$ is the identity on $\mathcal{N}$. This completes the induction procedure and thus the proposition is proved.

Proposition 3: Let

$$\mathcal{C} = \mathcal{A} \underset{\tau}{\otimes} \mathcal{B}\,, \qquad \mathcal{P} = \mathcal{A} \underset{\sigma}{\otimes} \mathcal{N}$$

be twisted products with twisted parts of the twisted differentials d_τ and d_σ respectively in which $\mathcal{N}$ is the minimal model of $\mathcal{B}$. If some DGA-morphism

$$\tilde{\rho} : \mathcal{P} \to \mathcal{C}$$

respects the twisted differentials with ident. $\mathcal{A} \to \mathcal{A}$ as the induction and some DGA-morphism $\rho_{\mathcal{B}} : \mathcal{N} \to \mathcal{B}$ as the restriction, then $\tilde{\rho}$ is an H-isomorphism if and only if $\rho_{\mathcal{B}}$ is so, or $\rho_{\mathcal{B}}$ is a minimal morphism associated to $\mathcal{B}$.

Proof: Write any element $z \in \mathcal{C}$ in the form

$$z = z_0 + z_1 + \dots + z_k$$

in which each z_p is a linear combination of terms of the form Xb with $b \in \mathcal{B}$ and $X \in \mathcal{A}_p$ of degree p in $\mathcal{A}$. Provisionally, we call such elements of rank $\geq p$ if $z_0 = \dots = z_{p-1} = 0$. The collection of all elements in $\mathcal{C}$ of rank $\geq p$ then forms a sub-DGA of $\mathcal{C}$, to be denoted by $\tilde{\mathcal{C}}^{(p)}$. Similarly, form the sub-DGA $\tilde{\mathcal{P}}^{(p)}$ of $\mathcal{P}$ for all $p \geq 0$. Then $\tilde{\rho} : \mathcal{P} \to \mathcal{C}$, when restricted to elements of $\tilde{\mathcal{P}}^{(p)}$, will clearly induce DGA-morphisms

$$\tilde{\rho}^{(p)} : \tilde{\mathcal{P}}^{(p)} \to \tilde{\mathcal{C}}^{(p)}\,.$$

Now, define a filtration in $\mathcal{C}$ by setting

$$F^p\mathcal{C} = \tilde{\mathcal{C}}^{(p)}$$

so that

$$F^0\mathcal{C} = \mathcal{C} \supset \tilde{\mathcal{C}}^{(1)} \supset \dots \supset \tilde{\mathcal{C}}^{(p)} \supset \dots$$

Then, for the induced spectral sequence, we have

$$E_0^p \approx \mathcal{A}_p \otimes \mathcal{B}\,, \qquad d_0 = d_{\mathcal{B}} \mid \mathcal{B}\,,$$

$$E_1^p \approx \mathcal{A}_p \underset{\tau}{\otimes} H(\mathcal{B})$$

with d_1 in E_1 a twisted differential induced from that of $\mathcal{C}$.

Similarly, we define a filtration in $\mathcal{P}$ by setting

$$F^p\mathcal{P} = \tilde{\mathcal{P}}^{(p)} ,$$

Then we have

$$F^0\mathcal{P} = \mathcal{P} \supset \tilde{\mathcal{P}}^{(1)} \supset \ldots \supset \tilde{\mathcal{P}}^{(p)} \supset \ldots ,$$

$$E_0^p \approx \mathcal{A}_p \otimes \mathcal{N} , \quad d_0 = d_{\mathcal{N}} | \mathcal{N} ,$$

$$E_1^p \approx \mathcal{A}_p \underset{\sigma}{\otimes} H(\mathcal{N}) ,$$

with d_1 a twisted differential induced from that of $\mathcal{P}$.

Now, $\tilde{\rho} : \mathcal{P} \to \mathcal{C}$ clearly respects the filtration. If $\rho_{\mathcal{B}}$ is a minimal morphism then $\tilde{\rho}$ will induce an isomorphism

$$E_1^p\mathcal{P} \approx E_1^p\mathcal{C} .$$

By comparison theorems of spectral sequences we therefore have

$$\tilde{\rho}_H : H(\mathcal{P}) \approx H(\mathcal{C}) ,$$

or $\tilde{\rho}$ is an H-isomorphism. The converse follows also from the comparison theorems.

Proposition 4: Let $\mathcal{C} = \mathcal{A} \underset{\tau}{\otimes} \mathcal{B}$ be a twisted product and $\mathcal{N}$ be the minimal model of $\mathcal{B}$. Then, in the tensor product $\mathcal{A} \otimes \mathcal{N}$, we can introduce some twisted differential with twisted part d_σ such that, for the resulting twisted product $\mathcal{A} \underset{\sigma}{\otimes} \mathcal{N}$, there is some DGA-morphism $\tilde{\rho} : \mathcal{A} \underset{\sigma}{\otimes} \mathcal{N} \to \mathcal{A} \underset{\tau}{\otimes} \mathcal{B}$, respecting the twisting differentials with the identity on $\mathcal{A}$ and some minimal morphism $\rho_{\mathcal{B}} : \mathcal{N} \to \mathcal{B}$ as the respective induction and restriction.

Proof: Consider the natural inclusion $i_{\mathcal{A}} : \mathcal{A} \subset \mathcal{A} \underset{\tau}{\otimes} \mathcal{B}$. Following the method described in the proof of the existence theorem of minimal models in §II 3, we can successively adjoin new generators to $\mathcal{A}$ to form a DGA $\mathcal{A} \underset{\sigma}{\otimes} \mathcal{N}'$, with $\mathcal{N}'$ a minimal DGA. During the process we also extend $i_{\mathcal{A}}$ successively to get a DGA-morphism $\tilde{\rho} : \mathcal{A} \underset{\sigma}{\otimes} \mathcal{N}' \to \mathcal{A} \underset{\tau}{\otimes} \mathcal{B}$, which is an H-isomorphism. The present proposition then follows from Prop. 3, with $\mathcal{N}'$ a minimal model of $\mathcal{B}$, to be identified with the given one $\mathcal{N}$ by §II 4.

From the preceding propositions we now deduce the following theorem which has the origin in works of J.C. Moore, cf. Sem. Cartan 1954/55, Exp. 3.

<u>Theorem</u>: Let $\mathcal{A} \underset{\tau}{\otimes} \mathcal{B}$ be a twisted tensor product and $\mathcal{M}, \mathcal{N}$ be minimal models of $\mathcal{A}, \mathcal{B}$. Then, there is a twisted tensor product $\mathcal{M} \underset{\sigma}{\otimes} \mathcal{N}$ and a DGA-morphism $\tilde{\rho} : \mathcal{M} \underset{\sigma}{\otimes} \mathcal{N} \to \mathcal{A} \underset{\tau}{\otimes} \mathcal{B}$ respecting the twisted differentials, such that the induction $\rho_{\mathcal{A}} : \mathcal{M} \to \mathcal{A}$ and the restriction $\rho_{\mathcal{B}} : \mathcal{N} \to \mathcal{B}$ are respective minimal morphisms and $\tilde{\rho}$ itself is an H-isomorphism.

Chapter III

THE DE RHAM - SULLIVAN THEOREM AND I^*-MEASURE

III.1 THE DE RHAM - SULLIVAN ALGEBRA OF A SIMPLICIAL COMPLEX AND THE DE RHAM - SULLIVAN THEOREM

Let K be a connected countable simplicial complex in weak topology with vertices arranged in a definite order

$$v_0 < v_1 < \ldots < v_i < \ldots .$$

A set of indices $J = (j_0, j_1, \ldots, j_p)$ will then be said to be in the normal order if $j_0 < j_1 < \ldots < j_p$ and $\sigma_J = (v_{j_0} v_{j_1} \ldots v_{j_p})$ is a simplex of K. The p-simplex σ_J is then also said to be normally-ordered. To each vertex v_i let us define now a function t_i on the space $|K|$ of K, as follows. For v_i let K_i be the closed subcomplex of K consisting of all simplexes with v_i as one of its vertices as well as all their faces. For a point p of $|K|$ not in $|K_i|$ the value of t_i at p will be defined to be 0. For a point $p \in |K|$ lying in certain simplex σ of K_i with one of its vertex v_i, t_i at p will be defined to be the barycentric coordinate of p in that simplex corresponding to v_i. The set of functions t_i on $|K|$ is then point-finite in the following sense. For each point $p \in |K|$ which lies in the interior of the simplex of K uniquely determined by the point p, the only functions non-zero at p are just those corresponding to the vertices of that simplex. It follows in particular that for any point $p \in |K|$, we have

$$\Sigma\, t_i(p) = 1 .$$

The summation over all indices is i which may be infinite in number but has, in fact, only finite number of non-zero terms.

For each vertex v_i let us also associate formally a differential dt_i. Now, consider any normally ordered p-simplex $\sigma_J = (v_{j_0} v_{j_1} \ldots v_{j_p})$ of K with vertices $v_{j_0}, v_{j_1}, \ldots, v_{j_p}$ where $J = (j_0, j_1, \ldots, j_p)$ is a normally ordered (p+1)-tuple of indices with $j_0 < j_1 < \ldots < j_p$. For such a σ_J we shall denote by $\bar{A}^s(\sigma_J)$, $s \geq 0$, the set of all formal differential forms of degree s of the form

$$\sum_H F_H dt_H = \sum_H F_H dt_{h_1} \ldots dt_{h_s}$$

in which $\sum_H$ runs over all ordered sub-s-tuples $H = (h_1, \ldots, h_s)$ from J, dt_H stands for $dt_{h_1} \ldots dt_{h_s}$, and F_H are polynomials in t_j, $j \in J$ with coefficients in a fixed field $\underline{k}$ of characteristic 0. The direct sum of all these $\bar{A}^s(\sigma_J)$ for $s \geq 0$, viz.

$$\bar{A}^*(\sigma_J) = \sum_s \bar{A}^s(\sigma_J)$$

will then form a DGA over $\underline{k}$ with the algebraic operations, exterior differentiation and gradation, all defined in the usual manner. Denote also by $\bar{B}^*(\sigma_J)$ the ideal of $\bar{A}^*(\sigma_J)$ generated by the forms

$$T_J = t_{j_0} + t_{j_1} + \dots + t_{j_p} - 1$$

and

$$dT_J = dt_{j_0} + dt_{j_1} + \dots + dt_{j_q} .$$

Then, the DGA

$$\bar{A}^*(\sigma_J)/\bar{B}^*(\sigma_J) = A^*(\sigma_J) = \Sigma A^s(\sigma_J)$$

will be called the <u>algebra of differential forms</u> on σ_J. For abuse of notations when no misunderstanding can occur, an element expressed as

$$w = \Sigma F_H dt_H$$

in $\bar{A}^*(\sigma_J)$ will also be used to denote its class in $A^*(\sigma_J)$. Sometimes it is also called a <u>formal expression</u> of its class.

Let J' be a subset of J and $\sigma_{J'}$ the face of σ_J spanned by vertices $v_{j'}$, with $j' \in J' \subset J$. Then the relations

$$t_{j''} = 0 , \quad dt_{j''} = 0 \quad (j'' \in J - J')$$

will induce a DGA-morphism of $\bar{A}^*(\sigma_J)$ to $\bar{A}^*(\sigma_{J'})$ which passes to a DGA-morphism of $A^*(\sigma_J)$ to $A^*(\sigma_{J'})$, to be called the <u>restriction</u> from σ_J to $\sigma_{J'}$. Now, we lay down the following fundamental definitions.

<u>Definition</u>: A collection of differential forms

$$w_J \in A^s_{\underline{k}}(\sigma_J)$$

of same degree s on each simplex σ_J of K will be said to be <u>compatible</u> if for each pair of simplexes $\sigma_{J'}$, σ_J of K with $\sigma_{J'}$ a face of σ_J the restriction of w_J on σ_J to $\sigma_{J'}$ is just $w_{J'}$. Such a compatible collection will be called a <u>differential form</u> of <u>degree</u> s <u>on K</u>. Denote by $A^s(K)$ the set of all such differential forms of K and by $A^*(K)$ the direct sum of these $A^s(K)$, $s \geq 0$. Then, under natural algebraic operations and exterior differentiation $A^*(K)$ will become a DGA over $\underline{k}$ which will be called the <u>deRham-Sullivan Algebra</u> of K (on the field $\underline{k}$).

Definition: The minimal model of the deRham-Sullivan algebra of K over $\underline{k}$ will be called the I^*-measure of K over $\underline{k}$ to be denoted by $I^*_{\underline{k}}(K)$:

$$I^*(K) \approx \min A^*(K) .$$

Remark: We have dropped $\underline{k}$ in all the notations. Thus, instead of $A^*_{\underline{k}}(K)$, $I^*_{\underline{k}}(K)$, etc. we have written simply $A^*(K)$, $I^*(K)$, etc. whenever no misunderstanding can occur. Similarly for $H^*(K) = H^*_{\underline{k}}(K)$, etc.

Notation: For a differential form $w \in A^*(K)$ the form in w on any simplex $\sigma \in K$ will be denoted by $w(\sigma)$. Moreover, for a form $\alpha \in A^*(\sigma)$, the restriction of α to a face σ' of σ will be denoted by $\alpha|\sigma'$. Similarly, for K' a subcomplex of K and $w \in A^*(K)$, the collection of all $w(\sigma')$ with $\sigma' \in K'$ is also compatible and forms a differential form on K'. This differential form, to be called <u>restriction</u> of w on K', will sometimes be denoted by $w|K'$.

The aim of this chapter is now to prove the following fundamental

Theorem of deRham-Sullivan: The homology of the I^*-measure of a simplicial complex K is isomorphic as graded algebras to the cohomology-ring-measure of K on coefficient field $\underline{k}$:

$$H(I^*(K)) \approx H^*(K) .$$

We shall postpone the proof of this theorem to §III 3 which is based on some work of A. Weil [1].

III.2 THE WEIL DGA OF A COMPLEX

For the proof of the deRham-Sullivan Theorem as stated in §III 1 to be given later, we shall first make some preparations in the present section.

Let K be accordingly a connected countable simplicial complex in weak topology with $A^*(K)$ its deRham-Sullivan algebra and $I^*(K)$ its I^*-measure.

In order to prove the theorem, viz

$$H(I^*(K)) \approx H^*(K) ,$$

it is clear that it will be enough to prove

$$H(A^*(K)) \approx H^*(K) .$$

If we denote by $C^*(K) = \Sigma C^n(K)$ the usual cochain-ring-measure of K on coefficient field $\underline{k}$, then this is further equivalent to prove

$$H(A^*(K)) \approx H(C^*(K)) .$$

Remark that $C^*(K)$ possesses a non-commutative DGA structure under the Alexander-Whitney multiplication, and the isomorphism to be proved is an algebraic one. Note that we have written simply $C^*(K)$ and $C^n(K)$ instead of $C^*_{\underline{k}}(K)$, $C^n_{\underline{k}}(K)$ with $\underline{k}$ dropped.

The proof to be given in the next section will be based on one due to Weil in the case of differential manifolds with due modifications. Roughly speaking, Weil introduced some non-commutative DGA W^* and some DGA-morphisms d and δ as in the diagram below:

$$(D) \qquad C^*(K) \xrightarrow{d} W^* \xleftarrow{\delta} A^*(K) .$$

The morphisms d and δ are then shown to induce algebraic isomorphisms

$$(D)_H \qquad H(C^*(K)) \underset{\approx}{\xrightarrow{d_H}} H(W^*) \underset{\approx}{\xleftarrow{\delta_H}} H(A^*(K))$$

which proves the theorem in question.

To define the DGA W^* let us first introduce a collection of subcomplexes of K as follows. Denote the totality of indices $(1, 2, \ldots, n, \ldots)$ by N and the totality of all $(p+1)$-tuples of indices $J = (j_0, j_1, \ldots, j_p)$ in normal order with $j_0 < j_1 < \ldots < j_p$ by N_p. For each $J \in N_p$ let K_J be now the closed subcomplex of K, consisting of all simplexes with σ_J as its face as well as all their faces. In particular we set $K_{(i)} = K_i$. For $J' \subset J$, J, J' both in normal order, K_J is a subcomplex of $K_{J'}$ and the inclusion will induce a DGA-morphism $A^*(K_{J'}) \to A^*(K_J)$ which consists in restricting each compatible collection of differential forms on simplexes of $K_{J'}$, to a collection of those on simplexes of K_J only. Similarly, for any J in normal order K_J is a subcomplex of K and the natural inclusion will induce a DGA-morphism by restriction $A^*(K) \to A^*(K_J)$. For any $w \in A^*(K)$ or $A^*(K_{J'})$, its restriction in $A^*(K_J)$ will be simply denoted by $w|K_J$.

Let $p \geq 0$, $m \geq 0$ be fixed integers. For each normally-ordered p-simplex σ_J of K, $J \in N_p$, let us take a differential form of degree m on K_J : $w_J \in A^m(K_J)$.

The set of all such collections under natural algebraic operations

$$w = (w_J)$$

will form some $\underline{k}$-module $W^{m,p}$. Remark that w_J is itself a (compatible) collection of differential forms on each simplex in the complex K_J.

Convention: For a permutation J' of the normally-ordered subset J with $\sigma_J \in K$, we shall set for convenience

$$w_{J'} = \varepsilon w_J \in A^*(K_J)$$

in which $\varepsilon = +1$ or -1 according to the permutation being even or odd.

Now, form the direct sum

$$W^{**} = \sum W^{m,p}$$

which will become a bi-graded algebra on K with a multiplication (not-anticommutative)

$$W^{m_1,p_1} \cdot W^{m_2,p_2} \subset W^{m_1+m_2,p_1+p_2}$$

defined as follows.

Let $w' \in W^{m_1,p_1}$, $w'' \in W^{m_2,p_2}$. For each normally-ordered (p_1+p_2)-simplex σ_J of K, with $J = (j_0, j_1, \ldots, j_{p_1+p_2})$ in normal order, consider the normally ordered p_1-simplex $\sigma_{J'}$, $J' = (j_0, \ldots, j_{p_1})$, and the normally ordered p_2-simplex $\sigma_{J''}$, $J'' = (j_{p_1}, \ldots, j_{p_1+p_2})$. With respect to the Alexander-Whitney product we then have

$$\sigma_{J'} \cdot \sigma_{J''} = \sigma_J .$$

Let $w'_{J'} \in A^{m_1}(K_{J'})$ be the form on $K_{J'}$ in the collection $w' \in W^{m_1,p_1}$. Similarly, for $w''_{J''} \in A^{m_2}(K_{J''})$ in $w'' \in W^{m_2,p_2}$. By definition we now set

$$w_J = w'_{J'J}\, w''_{J''J} \in A^{m_1+m_2}(K_J) ,$$

in which

$$w'_{J'J} = w'_{J'} \mid K_J , \quad w''_{J''J} = w''_{J''} \mid K_J .$$

With J running over all $(p_1 - p_2)$-tuples of indices in normal order, we then get a collection

$$w = (w_J) \in W^{m_1+m_2,p_1+p_2}$$

which is defined to be the product $w'w''$.

Next, introduce in W^{**} some differentials

$$d : W^{m,p} \to W^{m+1,p} \quad \text{and}$$

$$\delta : W^{m,p} \to W^{m,p+1}$$

as follows.

For the definition of d let us consider any differential form $w = (w_J) \in W^{m,p}$ with $w_J \in A^m(K_J)$, $J \in N_p$. We then simply set by definition

$$dw = (dw_J) \in W^{m+1,p} .$$

To define $\delta : W^{m,p} \to W^{m,p+1}$ let us consider $w = (w_J) \in W^{m,p}$ as above. For each (p+1)-simplex σ_H of K with $H = (h_0, \dots, h_{p+1}) \in N_{p+1}$, let $H_\nu = (h_0, \dots, \hat{h}_\nu, \dots, h_{p+1}) \in N_p$. Then, we put by definition

$$(1) \qquad \eta_H = \Sigma\, (-1)^\nu \cdot w_{H_\nu H} \in A^m(K_H) ,$$

in which

$$w_{H_\nu H} = w_{H_\nu} \mid K_H .$$

We then define

$$\delta w = (\eta_H) \in W^{m,p+1}$$

with H running over N_{p+1}.

<u>Remark</u>: Under the convention given before we verify easily that the formula

$$(1) \qquad \eta_H = \Sigma\, (-1)^\nu \cdot w_{H_\nu H}$$

remains true for (p+2)-tuples of indices H not in normal order, but a permutation of a (p+2)-tuple in normal order.

Now, we put

$$W^* = \sum_{q \geq 0} W^q$$

$$W^q = \sum_{p=0}^{q} W^{q-p,p}$$

$$W^{m,*} = \sum_{p \geq 0} W^{m,p}$$

$$W^{*,p} = \sum_{m \geq 0} W^{m,p} .$$

It is easy to verify that both d and δ are differentials in $W^{*,p}$ resp. $W^{m,*}$ for any $p \geq 0$ or $m \geq 0$. Moreover, we have also the following

Lemma 1: d and δ are commutative, i.e.

$$d\delta = \delta d : W^{m,p} \to W^{m+1,\, p+1}$$

for any $m \geq 0$, $p \geq 0$.

Proof: This is evident since from (1) we get by differentiation the formula with w, η replaced by dw and $d\eta$.

From the lemma it follows that in defining the gradation in W^* by $\deg w = q$ for $w \in W^q$, then

$$(2) \qquad D = d + \bar{\delta} : W^q \to W^{q+1}$$

where

$$(3) \qquad \bar{\delta} = (-1)^m \delta \mid W^{m,*}$$

will be a differential in W^*.

Definition: The DGA W^* with differential D will be called the Weil DGA of the complex K.

Remark: The multiplication in the Weil DGA is not anticommutative. On the other hand, it is bi-graded with bi-degree (m,p) for elements in $W^{m,p}$ and the sub-DGA $W^{*,0}$ is anti-commutative, so that it is the commutative DGA in the usual sense with d as the differential.

In what follows we shall define DGA-morphisms d on $C^*(K)$ and δ on $A^*(K)$ to make the diagram (D^W) below commutative:

$$
(D^W)\qquad
\begin{array}{ccccccccccccc}
 & & \vdots & & \vdots & & & & \vdots & & \vdots & & \\
 & & \uparrow{\scriptstyle\delta} & & \uparrow{\scriptstyle\delta} & & & & \uparrow{\scriptstyle\delta} & & \uparrow{\scriptstyle\delta} & & \\
0 & \longrightarrow & C^{p+1}(K) & \xrightarrow{d} & W^{0,p+1} & \xrightarrow{d} & \cdots & \xrightarrow{d} & W^{m,p+1} & \xrightarrow{d} & W^{m+1,p+1} & \xrightarrow{d} & \cdots \\
 & & \uparrow{\scriptstyle\delta} & & \uparrow{\scriptstyle\delta} & & & & \uparrow{\scriptstyle\delta} & & \uparrow{\scriptstyle\delta} & & \\
0 & \longrightarrow & C^{p}(K) & \xrightarrow{d} & W^{0,p} & \xrightarrow{d} & \cdots & \xrightarrow{d} & W^{m,p} & \xrightarrow{d} & W^{m+1,p} & \xrightarrow{d} & \cdots \\
 & & \uparrow{\scriptstyle\delta} & & \uparrow{\scriptstyle\delta} & & & & \uparrow{\scriptstyle\delta} & & \uparrow{\scriptstyle\delta} & & \\
 & & \vdots & & \vdots & & & & \vdots & & \vdots & & \\
 & & \uparrow{\scriptstyle\delta} & & \uparrow{\scriptstyle\delta} & & & & \uparrow{\scriptstyle\delta} & & \uparrow{\scriptstyle\delta} & & \\
0 & \longrightarrow & C^{0}(K) & \xrightarrow{d} & W^{0,0} & \xrightarrow{d} & \cdots & \xrightarrow{d} & W^{m,0} & \xrightarrow{d} & W^{m+1,0} & \xrightarrow{d} & \cdots \\
 & & \uparrow & & \uparrow{\scriptstyle\delta} & & & & \uparrow{\scriptstyle\delta} & & \uparrow{\scriptstyle\delta} & & \\
 & & 0 & \longrightarrow & A^{0}(K) & \xrightarrow{d} & \cdots & \xrightarrow{d} & A^{m}(K) & \xrightarrow{d} & A^{m+1}(K) & \xrightarrow{d} & \cdots \\
 & & & & \uparrow & & & & \uparrow & & \uparrow & & \\
 & & & & 0 & & & & 0 & & 0 & &
\end{array}
$$

For the definition of d on $C^*(K)$ let $\gamma \in C^p(K)$ be given. For any $(p+1)$-tuple of indices $H = (h_0, \ldots, h_p)$ in normal order let $\gamma_H \in A^0(K_H)$ be the differential form of degree 0 on K_H which takes on each simplex of K_H the constant-valued polynomial $\gamma(\sigma_H)$. Then, we put as definition

$$d(\gamma) = (\gamma_H) \in W^{0,p}$$

in which H runs over N_p. The d thus defined gives morphisms

$$d : C^*(K) \to W^*$$

with

$$dC^p(K) \subset W^{0,p} .$$

<u>Lemma 2</u>: The morphism $d : C^*(K) \to W^*$ defined above commutes with differentials δ in $C^*(K)$ and W^* and is a DGA-morphism.

<u>Proof</u>: Clear from the very definitions of the various morphisms involved.

Now, for any $w \in A^m(K)$ let

$$w_j = w|K_j \in A^m(K_j)$$

for any $j \in N$. Then, (w_j) with j running over all $j \in N$ will be defined as

$$\delta w = (w_j) \in W^{m,0} .$$

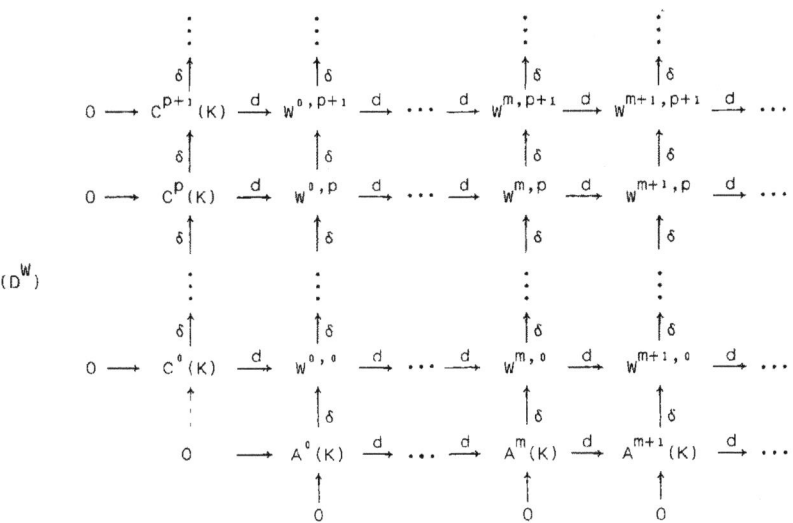

Lemma 3: The morphism

$$\delta : A^*(K) \to W^*$$

defined above commutes with the differentials d in $A^*(K)$ and W^* and is a DGA-morphism.

Proof: It is also clear from the very definitions. Remark that $\delta A^*(K) \subset W^{*,0}$ is anti-commutative though W^* is not.

It can be shown that the sequence formed from each row of (D^W)

$$0 \longrightarrow C^p(K) \xrightarrow{d} W^{0,p} \xrightarrow{d} \dots \xrightarrow{d} W^{m,p} \xrightarrow{d} \dots$$

and the sequence formed from each column of (D^W)

$$0 \longrightarrow A^m(K) \xrightarrow{\delta} W^{m,0} \xrightarrow{\delta} \dots \xrightarrow{\delta} W^{m,p} \xrightarrow{\delta} \dots$$

all are exact. From these it would follow the deRham-Sullivan theorem by a reasoning as given in Bott [1]. However, the exactness of the above sequences are not evident and require some arguments not entirely trivial. Moreover, such proofs are only existential and non-constructive in character. So we shall avoid to do it in this way and, on the contrary, we shall follow the original path of Weil's paper in proving in a direct and constructive manner that in (D) both d and δ are H-isomorphisms. This will complete the proof of the deRham-Sullivan Theorem and will be given in the next section.

III.3 PROOF OF THE DE RHAM – SULLIVAN THEOREM

The proof of the theorem in question is based on the diagram

$$(D) \qquad C^*(K) \xrightarrow{d} W^* \xleftarrow{\delta} A^*(K)$$

already defined in the preceding section, which has been completed to a commutative diagram (D^W). Now, we shall introduce further morphisms I,K into (D^W) to get one below:

$$
(D^W)\qquad
\begin{array}{ccccccccc}
 & \vdots & & \vdots & & \vdots & & \vdots & \\
 & \delta\uparrow & & \delta\uparrow\downarrow K & & \delta\uparrow\downarrow K & & \delta\uparrow\downarrow K & \\
0 \longrightarrow & C^{p+1}(K) & \underset{I}{\overset{d}{\rightleftarrows}} & W^{0,p+1} & \rightleftarrows \cdots \rightleftarrows & W^{m,p+1} & \underset{I}{\overset{d}{\rightleftarrows}} & W^{m+1,p+1} & \rightleftarrows \cdots \\
 & \delta\uparrow & & \delta\uparrow\downarrow K & & \delta\uparrow\downarrow K & & \delta\uparrow\downarrow K & \\
0 \longrightarrow & C^{p}(K) & \underset{I}{\overset{d}{\rightleftarrows}} & W^{0,p} & \rightleftarrows \cdots \rightleftarrows & W^{m,p} & \underset{I}{\overset{d}{\rightleftarrows}} & W^{m+1,p} & \rightleftarrows \cdots \\
 & \delta\uparrow & & \delta\uparrow\downarrow K & & \delta\uparrow\downarrow K & & \delta\uparrow\downarrow K & \\
 & \vdots & & \vdots & & \vdots & & \vdots & \\
 & \delta\uparrow & & \delta\uparrow\downarrow K & & \delta\uparrow\downarrow K & & \delta\uparrow\downarrow K & \\
0 \longrightarrow & C^{0}(K) & \underset{I}{\overset{d}{\rightleftarrows}} & W^{0,0} & \rightleftarrows \cdots \rightleftarrows & W^{m,0} & \underset{I}{\overset{d}{\rightleftarrows}} & W^{m+1,0} & \rightleftarrows \cdots \\
 & \uparrow & & \delta\uparrow\downarrow K & & \delta\uparrow\downarrow K & & \delta\uparrow\downarrow K & \\
 & 0 \longrightarrow & & A^{0}(K) & \xrightarrow[d]{} \cdots \longrightarrow & A^{m}(K) & \xrightarrow[d]{} & A^{m+1}(K) & \longrightarrow \cdots \\
 & & & \uparrow & & \uparrow & & \uparrow & \\
 & & & 0 & & 0 & & 0 &
\end{array}
$$

In this diagram the morphisms I, K are defined as follows.

<u>Definition</u> of $I : W^{0,p} \to C^p(K)$.

For any $f = (f_J) \in W^{0,p}$ with J running over all normally ordered $(p+1)$-tuples $(j_0, \ldots, j_p) \in N_p$ and $f_J \in A^0(K_J)$ we just define $I(f) \in C^p(K)$ as the cochain which takes the value $f_J((v_{j_0}))$ on the simplex σ_J.

<u>Lemma 1</u>: $Id = \text{ident.} : C^p(K) \to C^p(K)$.

<u>Proof</u>: This follows directly from the definitions of I and d.

<u>Definition</u> of $I : W^{m+1,p} \to W^{m,p}$, $m \geq 0$.

Consider any normally-ordered p-simplex σ_J of K, with $J = (j_0, \ldots, j_p)$ in normal order. Let τ be any simplex in the complex K_J. We shall first define a morphism

$$I_{J,\tau} : A^{m+1}(\tau) \to A^m(\tau)$$

as follows.

First suppose that τ has v_{j_0} as a vertex with vertices $v_{k_1}, \ldots, v_{k_q}$ besides v_{j_0}, $k_1 < \ldots < k_q$. For the barycentric-coordinate functions in τ we have

$$t_{j_0} = 1 - \sum_{s=1}^{q} t_{k_s} ,$$

$$dt_{j_0} = - \sum_{s=1}^{q} dt_{k_s} .$$

Any form $\alpha \in A^{m+1}(\tau)$ has consequently a unique representative in the form

$$\alpha = \sum_H F_{h_0 \ldots h_m} dt_{h_0} \ldots dt_{h_m} = \sum F_H dt_H ,$$

in which $\sum_H$ runs over all normally-ordered $H = (h_0, \ldots, h_m) \subset (k_1, \ldots, k_q)$ and $F_H = F_{h_0 \ldots h_m}$ are polynomials in $t_{k_1}, \ldots, t_{k_q}$ with coefficients in $\underline{k}$. Now, for any polynomial F in $t_{k_1}, \ldots, t_{k_q}$ let us write it in the form

$$F = \sum_{i \geq 0} F^{(i)}$$

in which $F^{(i)}$ is the homogeneous part of F of degree i. We set then by definition

$$I_{J,\tau}\alpha = \sum_{H,\nu,i} \left[(-1)^{\nu} \cdot \frac{1}{m+i+1} \cdot F_H^{(i)} \cdot t_{h_\nu} dt_{H_\nu} \right] \in A^m(\tau) ,$$

in which $H_\nu = (h_0, \ldots, \hat{h}_\nu, \ldots, h_m)$.

Now, consider an element

$$w = (w_J) \in W^{m+1,p} .$$

For each normally-ordered $\sigma_J \in K$ of dimension p, $w_J \in A^{m+1}(K_J)$ is a compatible collection of differential forms $w_J(\tau) \in A^{m+1}(\tau)$ with $\tau \in K_J$. Let $J = (j_0, \ldots, j_p)$ in normal order, then for any τ having v_{j_0} as a vertex we shall set $I_J w_J(\tau) = I_{J,\tau} w_J(\tau) \in A^m(\tau)$ as defined above. These forms are clearly compatible and can be extended uniquely to a compatible collection of differential forms for all $\tau \in K_J$ which will then be denoted by $I_J w_J \in A^m(K_J)$. We set then by definition

$$I(w) = (I_J w_J) \in W^{m,p} .$$

<u>Lemma 2</u>: For the morphisms I we have $(m, \; p \geq 0)$

$$dI + Id = \text{ident.} \mid W^{m,p} .$$

<u>Proof</u>: Let us consider first the case $m = 0$. Suppose thus, given an element

$$f = (f_J) \in W^{0,p} ,$$

in which J runs over all normally-ordered $(p+1)$-tuples $(j_0, \ldots, j_p) \in N_p$ and $f_J \in A^0(K_J)$ is a compatible collection of polynomials on simplexes of K_J. Let

$$\begin{aligned} I(f) &= \gamma \in C^p(K) , \\ dI(f) &= d\gamma = g \in W^{0,p} , \\ d(f) &= w \in W^{1,p} , \\ Id(f) &= Iw = h \in W^{0,p} . \end{aligned}$$

Then we have to prove

$$g + h = f \in W^{0,p} .$$

For this purpose let $J = (j_0, \ldots, j_p)$ be as above and τ be any simplex in K_J with one vertex at v_{j_0}, the other vertices being $v_{k_1}, \ldots, v_{k_s}$ $(k_1 < \ldots < k_s)$ which will span with σ_J a simplex of K_J. Let F be $f_J(\tau)$ on τ with t_{j_0} already replaced by $1 - \sum_\nu t_{k_\nu}$. Write F in the form

$$F = \sum_{i \geq 0} F^{(i)}$$

in which $F^{(i)}$ is the homogeneous part of degree i of F. Then

$$\begin{aligned} w_J(\tau) &= df_J(\tau) = dF \\ &= \sum_{\nu, i>0} \frac{\partial F^{(i)}}{\partial t_{k_\nu}} dt_{k_\nu} , \\ h_J(\tau) &= I_{J,\tau}\, w_J(\tau) = \sum_{\nu, i>0} \frac{1}{i} \frac{\partial F^{(i)}}{\partial t_{k_\nu}} \cdot t_{k_\nu} \\ &= \sum_{i>0} F^{(i)} = F - F^{(0)} \\ &= f_J(\tau) - f_J(v_{j_0}) . \end{aligned}$$

On the other hand

$$\begin{aligned} \gamma(\sigma_J) &= f_J(v_{j_0}) , \\ g_J(\tau) &= (d\gamma)_J(\tau) = \gamma(\sigma_J) = f_J(v_{j_0}) . \end{aligned}$$

Hence,

$$g_J(\tau) + h_J(\tau) = f_J(\tau)$$

for $\tau \in K_J$ with one of the vertices of τ at v_{j_0}. This will then still be true for any $\tau \in K_J$. It follows that

$$g_J + h_J = f_J$$

in K_J. Since this is true for all p-simplexes σ_J, we have

$$g + h = f$$

as to be asserted.

Next, consider the case $m > 0$ or what is the same, the case of

$$dI + Id = \text{ident.} \mid W^{m+1,p}$$

for $m \geq 0$. To prove it, let us consider a normally-ordered (p+1)-tuple $J = (j_0, \dots, j_p)$ and a simplex $\tau \in K_J$ with one of its vertex at v_{j_0} and other vertices $v_{k_1}, \dots, v_{k_q}$ with $k_1 < \dots < k_q$ as before. Consider also any form

$$\alpha = \Sigma F_H dt_H = \Sigma F_H^{(i)} dt_H \in A^{m+1}(\tau)$$

with H running over all normally-ordered (m+1)-tuples $(h_0, \dots, h_m) \subset (k_1, \dots, k_q)$ and $F_H^{(i)}$ homogeneous of degree i in $t_{k_1}, \dots, t_{k_q}$ alone, the t_{j_0} already being replaced by $1 - \sum_\nu t_{k_\nu}$ and dt_{j_0} by $-\sum_\nu dt_{k_\nu}$. It is then clearly enough to prove that

$$(dI_{J,\tau} + I_{J,\tau} d)\alpha = \alpha .$$

By definition

$$I_{J,\tau}\alpha = \sum_{H,\nu,i} (-1)^\nu \cdot \frac{1}{m+1+i} \cdot F_H^{(i)} t_{h_\nu} dt_{H_\nu}$$

with $H_\nu = (h_0, \dots, \hat{h}_\nu, \dots, h_m)$. Whence,

$$dI_{J,\tau}\alpha = \Sigma_1 + \Sigma_2$$

where

$$\Sigma_1 = \sum_{H,i} \frac{m+1}{m+1+i} \cdot F_H^{(i)} dt_H ,$$

$$\Sigma_2 = \sum_{H,\mu,\nu,i} (-1)^\nu \cdot \frac{1}{m+1+i} \cdot \frac{\partial F_H^{(i)}}{\partial t_{k_\mu}} \cdot t_{h_\nu} dt_{k_\mu} dt_{H_\nu} .$$

On the other hand

$$d\alpha = \sum_{H,\mu,i} \frac{\partial F_H^{(i)}}{\partial t_{k_\mu}} \cdot dt_{k_\mu} dt_H .$$

Whence, by definition of $I_{J,\tau}$,

$$I_{J,\tau}\, d\alpha = \Sigma_3 + \Sigma_4 ,$$

in which

$$\Sigma_3 = \sum_{H,\mu,\nu,i} (-1)^{\nu+1} \cdot \frac{1}{(m+2)+(i-1)} \cdot \frac{\partial F_H^{(i)}}{\partial t_{k_\mu}} \cdot t_{h_\nu} dt_{k_\mu} dt_{H_\nu},$$

$$\Sigma_4 = \sum_{H,\mu,i} \frac{1}{(m+2)+(i-1} \cdot \frac{\partial F_H^{(i)}}{\partial t_{k_\mu}} \cdot t_{k_\mu} dt_H$$

$$= \sum_{H,i} \frac{i}{m+i+1} \cdot F_H^{(i)} dt_H ,$$

since $F_H^{(i)}$ is an homogeneous polynomial of degree i in t_{k_μ}.

Since $\Sigma_2 + \Sigma_3 = 0$ we see that

$$(dI_{J,\tau} + I_{J,\tau} d)\alpha = \Sigma_1 + \Sigma_4$$

$$= \Sigma F_H^{(i)} dt_H = \alpha ,$$

as to be proved.

Definition of $K : W^{m,0} \to A^m(K)$.

Let $w = (w_i) \in W^{m,0}$ with $w_i \in A^m(K_i)$, i running over all indices in N. Then $Kw \in A^m(K)$ will be defined to be the differential form such that for any normally-ordered p-simplex σ_J with $J = (j_0,\dots, j_p) \in N_p$,

$$(Kw)(\sigma_J) = \sum_\mu t_{j_\mu} w_{j_\mu} (\sigma_J) \in A^m(\sigma_J).$$

Lemma 3: $K\delta = \text{ident.} : A^m(K) \to A^m(K)$.

Proof: For $w \in A^m(K)$, $\delta w \in W^{m,0}$ is such that for any vertex v_i and any simplex $\sigma \in K_i$,

$$(\delta w)_i(\sigma) = w(\sigma) .$$

Hence, for any p-simplex σ_J with $J = (j_0,\dots, j_p)$ a normally-ordered (p+1)-tuple of indices, $K\delta w$ is by definition given by

$$
\begin{aligned}
(K\delta w)(\sigma_J) &= \sum_\mu t_{j_\mu} (\delta w)_{j_\mu}(\sigma_J) \\
&= \sum_\mu t_{j_\mu} w(\sigma_J) \\
&= w(\sigma_J) ,
\end{aligned}
$$

since $\Sigma_\mu t_{j_\mu} = 1$. Since σ_J is arbitrary we get $K\delta w = w$, as to be proved.

<u>Definition</u> of morphisms $K : W^{m,p+1} \to W^{m,p}$ $(m,p \geq 0)$.

Let

$$w = (w_J) \in W^{m,p+1}$$

be given with $w_J \in A^m(K_J)$ for all normally ordered σ_J of dimension p+1 in K. Consider any normally ordered p-simplex σ_H of K with $H = (h_0, \ldots, h_p) \in N_p$ normally ordered. Let v_k be a vertex of K which spans with σ_H a simplex $\sigma_{(kH)}$ of dimension p+1 in K, where (kH) is the set $(k, h_0, \ldots, h_p)$ not necessarily in normal order, cf. the convention in §III 2. For any simplex $\tau \in K_{(kH)}$ with v_k as a vertex, the form $t_k w_{(kH)}(\tau) \in A^m(\tau)$ is well-defined where $w_{(kH)}(\tau)$ is that form in the collection of forms $w_{(kH)} \in A^m(K_{(kH)})$. For any simplex $\tau \in K_H$ for which v_k is not a vertex we have clearly $t_k = 0$ on τ and we shall understand $t_k w_{(kH)}(\tau) \in A^m(\tau)$ simply as 0. The collection of forms $t_k w_{(kH)}(\tau)$ thus defined for all $\tau \in K_H$ is clearly a compatible one, and we shall denote it by $t_k w_{(kH)} \in A^m(K_H)$.

Now, we may define in K_H a differential form ζ_H by setting

$$\zeta_H = \Sigma\, t_k w_{(kH)} \in A^m(K_H) ,$$

in which the summation Σ is to be extended over all k not in H. It is well-defined since for any $\tau \in A^m(K_H)$ there are at most a finite number of non-zero terms in the summation corresponding to vertices of τ not of σ_H. We shall set by definition

$$Kw = (\zeta_H) \in W^{m,p} .$$

<u>Lemma 4</u>: For the morphisms K we have $(m,p \geq 0)$

$$\delta K + K\delta = \text{Ident.} \mid W^{m,p} .$$

Proof: First, consider the case $p = 0$. Suppose thus given

$$w = (w_i) \in W^{m,0}$$

in which $w_i \in A^m(K_i)$ with i running over all indices. Let

$$\delta w = \eta \in W^{m,1} ,$$
$$K\eta = \theta \in W^{m,0} ,$$
$$Kw = \zeta \in A^m(K) ,$$
$$\delta\zeta = \varphi \in W^{m,0} .$$

Then we have to prove that

$$\theta + \varphi = w \in W^{m,0} .$$

For this purpose let us consider any K_i and any simplex $\sigma \in K_i$ with one vertex at v_i and other vertices $v_{j_1}, \dots, v_{j_p}$. Then, we have for any $\nu = 1, \dots, p$ (cf. the convention in III 2),

$$\eta_{(j_\nu i)}(\sigma) = (\delta w)_{(j_\nu i)}(\sigma) = w_i(\sigma) - w_{j_\nu}(\sigma) ,$$
$$\theta_i(\sigma) = (K\eta)_i(\sigma) = \Sigma\, t_{j_\nu} \eta_{(j_\nu i)}(\sigma)$$
$$= \Sigma\, t_{j_\nu} w_i(\sigma) - \Sigma\, t_{j_\nu} w_{j_\nu}(\sigma) .$$

On the other hand

$$\zeta(\sigma) = (Kw)(\sigma) = t_i w_i(\sigma) + \Sigma\, t_{j_\nu} w_{j_\nu}(\sigma) ,$$
$$\varphi_i(\sigma) = (\delta\zeta)_i(\sigma) = \zeta(\sigma)$$
$$= t_i w_i(\sigma) + \Sigma\, t_{j_\nu} w_{j_\nu}(\sigma) .$$

It follows that

$$\theta_i(\sigma) + \varphi_i(\sigma) = (t_i + \Sigma\, t_{j_\nu}) w_i(\sigma) = w_i(\sigma) .$$

Since this should also be true for $\sigma \in K_i$ not having v_i as vertices we have

$$\theta_i + \varphi_i = w_i \quad \text{in} \quad K_i .$$

Since this is true for all i we have

$$\theta + \varphi = w$$

as to be proved.

Next consider the case $p > 0$. Suppose thus given a form $w \in W^{m,p}$ with

$$\eta = \delta w \in W^{m,p+1} ,$$
$$\theta = K\eta \in W^{m,p} ,$$
$$\zeta = Kw \in W^{m,p-1} ,$$
$$\varphi = \delta\zeta \in W^{m,p} .$$

For any normally-ordered (p+1)-tuple $J = (j_0,\ldots,j_p)$ with $J_\nu = (j_0,\ldots,\hat{j}_\nu,\ldots,j_p)$ we have now in K_J

$$\begin{aligned}
\varphi_J = (\delta\zeta)_J &= \Sigma\,(-1)^\nu \zeta_{J_\nu} \\
&= \Sigma\,(-1)^\nu (Kw)_{J_\nu} \\
&= \Sigma\,(-1)^\nu \sum_{k \notin J_\nu} t_k w_{(kJ_\nu)} \\
&= \Sigma\,(-1)^\nu \cdot [t_{j_\nu} w_{(j_\nu j_0 \ldots \hat{j}_\nu \ldots j_p)} + \sum_{k \notin J} t_k w_{(kJ_\nu)}] \\
&= \Sigma\, t_{j_\nu} w_J + \sum_{k \notin J} \sum_\nu (-1)^\nu \cdot t_k w_{(kJ_\nu)} .
\end{aligned}$$

On the other hand

$$\begin{aligned}
\theta_J = (K\eta)_J &= \sum_{k \notin J} t_k \eta_{(kJ)} \\
&= \sum_{k \notin J} t_k (\delta w)_{(kJ)} \\
&= \sum_{k \notin J} t_k w_J + \sum_{k \notin J} \sum_\nu (-1)^{\nu+1} t_k w_{(kJ_\nu)} .
\end{aligned}$$

It follows that

$$\theta_J + \varphi_J = \sum_{k \in N} t_k w_J = w_J .$$

Since this is true for all (p+1)-tuples J we have

$$\theta + \varphi = w ,$$

as to be asserted.

In W^* and $C^*(K)$ let us set

$$\bar{\delta} \mid W^{m,p} = (-1)^m \delta ,$$
$$\bar{\delta} \mid C^p(K) = -\delta .$$

Then

$$d\bar{\delta} = -\bar{\delta}d \mid W^* ,$$
$$d\bar{\delta} = -\bar{\delta}d \mid C^*(K) ,$$

and

$$D = d + \bar{\delta} \mid W^* .$$

Lemma 5: For $q \geq 0$ we have

$$(I\bar{\delta})^{q+1} d = I\bar{\delta} d (I\bar{\delta})^q : W^{q,p} \to W^{0,p+q+1} .$$

Proof: This is evident for $q = 0$. Proceed by induction we have on $W^{q,p}$,

$$(I\bar{\delta})^{q+1} d = -(I\bar{\delta})^q I d \bar{\delta} = -(I\bar{\delta})^q (1 - dI)\bar{\delta} = +(I\bar{\delta})^q dI\bar{\delta} ,$$

as $I\bar{\delta} W^{q,p} \in W^{q-1,p+1}$, so by induction,

$$(I\bar{\delta})^{q+1} d = I\bar{\delta} d (I\bar{\delta})^q ,$$

as to be proved.

Let us define now morphisms

$$d' : W^m \to C^m(K)$$

as follows. For any element

$$w = \sum_{q=0}^{m} w_q \in W^m \quad \text{with} \quad w_q \in W^{q,\, m-q} ,$$

let us set by definition

$$d'w = \Sigma\, I(-\bar{\delta} I)^q w_q \in C^m(K) .$$

Then we have

Lemma 6:

$$d'D = \delta d' \mid W^m .$$

Proof: Let $w = \Sigma w_q \in W^m$ as above with the convention $w_{-1} = 0$, $w_{m+1} = 0$. Then we have

$$Dw = \Sigma\, (d + \bar{\delta}) w_q = \Sigma\, v_q \in W^{m+1} ,$$

with

$$v_q = \bar{\delta} w_q + d w_{q-1} \in W^{q,m+1-q} .$$

Hence, by definition of d',

$$\begin{aligned} d'Dw &= \Sigma\, I(-\bar{\delta}I)^q\, v_q \\ &= \Sigma\, I(-\bar{\delta}I)^q\, (\bar{\delta}\, w_q + dw_{q-1}) \\ &= \Sigma\, I(-\bar{\delta}I)^q\, (\bar{\delta} - \bar{\delta}Id)w_q \,. \end{aligned}$$

By Lemmas 1,2,5 we have then

$$\begin{aligned} d'Dw &= \Sigma\, I(-\bar{\delta}I)^q\, \bar{\delta}dI\, w_q = \Sigma\, (-1)^q\, (I\bar{\delta})^{q+1}\, dIw_q \\ &= \Sigma\, (-1)^q\, (I\bar{\delta})^2\, d(I\bar{\delta})^{q-1}\, Iw_q \\ &= -\Sigma\, (-1)^q\, I\bar{\delta}Id\, \bar{\delta}(I\bar{\delta})^{q-1}\, Iw_q \\ &= -\Sigma\, (-1)^q\, I\bar{\delta}(1-dI)\bar{\delta}(I\bar{\delta})^{q-1}\, Iw_q \\ &= +\Sigma\, (-1)^q\, I\bar{\delta}d(I\bar{\delta})^q\, Iw_q \\ &= I\bar{\delta}d\ \Sigma\, I(-\bar{\delta}I)^q\, w_q \\ &= -Id\bar{\delta}d'w \,. \end{aligned}$$

Since $\bar{\delta}d'w \in C^{m+1}(K)$ and $Id = \text{ident.} \mid C^*(K)$, we finally get

$$d'Dw = \delta d'w \,.$$

<u>Proposition 1</u>: The DGA-morphism

$$d : C^*(K) \to W^*$$

is an H-isomorphism or

$$d_H : H(C^*(K)) \approx H(W^*) \,.$$

<u>Proof</u>: By Lemma 6, d' will induce a morphism

$$d'_H : H(W^*) \to H(C^*(K)) \,.$$

Since from definition we clearly have

$$d'd = \text{ident.} \mid C^*(K) \,,$$

we have also

$$d'_H d_H = \text{ident.}$$

so that d_H is in particular a monomorphism.

To prove that d_H is also an epimorphism, let $z = \Sigma z_q \in W^m$, $z_q \in W^{q,m-q}$ with $Dz = 0$. Then

$$\bar{\delta} z_q + d z_{q-1} = 0 .$$

Suppose that $z_k \neq 0$ while $z_q = 0$ for $q > k$. Then we have necessarily $d z_k = 0$ so that

$$\begin{aligned} z_k &= (dI + Id)\, z_k \\ &= dI\, z_k = DI\, z_k - \bar{\delta} I\, z_k \\ &\sim -\bar{\delta} I\, z_k \in W^{k-1,\, m-k+1} . \end{aligned}$$

By successive reductions we thus see that we have for some $\bar{z}_0 \in W^{0,m}$,

$$z \sim \bar{z}_0 \quad \text{in } W^* .$$

Since $D\bar{z}_0 = 0$ we have $d\bar{z}_0 = 0$ and $\delta\bar{z}_0 = 0$, so that

$$\bar{z}_0 = (Id + dI)\, \bar{z}_0 = dI\, \bar{z}_0$$

with

$$\begin{aligned} \delta I\, \bar{z}_0 &= Id\delta I\, \bar{z}_0 = I\delta dI\, \bar{z}_0 \\ &= I\delta\, \bar{z}_0 \\ &= 0 . \end{aligned}$$

It follows that $I\bar{z}_0$ is a cycle in $C^*(K)$ and the class of z or that of $\bar{z}_0 = dI\,\bar{z}_0$ is the image of the class of $I\,\bar{z}_0$ under d_H. This proves that d_H is also an epimorphism.

Entirely parallel to the above we also have now

Proposition 2: The DGA-morphism

$$\delta : A^*(K) \to W^*$$

is an H-isomorphism or

$$\delta_H : H(A^*(K)) \simeq H(W^*) .$$

Combining Propositions 1 and 2 the Theorem of deRham-Sullivan follows immediately. Moreover, from the proof of Prop. 1 we see that the isomorphism in the theorem will be induced by the explicit morphisms

$$d'\delta = I(-\bar{\delta}I)^m\delta = (-1)^{m(m+1)/2}\cdot(I\delta)^{m+1} ; A^m(K) \to C^m(K) .$$

It may also be induced by a similar explicit morphism in reverse direction, viz.

$$(-1)^{m(m+1)/2}\cdot(Kd)^{m+1} : C^m(K) \to A^m(K) .$$

III.4 INTEGRATION AND DUALITY

We have proved, for a connected countable simplicial complex K in weak topology, the deRham-Sullivan Theorem in the form of an algebraic isomorphism

$$H(A^*(K)) \approx H(C^*(K))$$

in which $C^*(K) = C^*_{\underline{k}}(K)$ with $\underline{k}$ a fixed field of characteristic 0 to be omitted throughout in the notations. Now, the cohomology-ring measure on the right-hand side, considered as a group, is known to be dual to the homology-group measure $H_\oplus(K) \approx H_\oplus^{\underline{k}}(K)$ of K on coefficient field $\underline{k}$ so that, as additive isomorphisms

$$H(A^*(K)) \approx \mathrm{Hom}_{\underline{k}}(H_\oplus(K), \underline{k})$$

$$\approx H(\mathrm{Hom}_{\underline{k}}(C(K), \underline{k}))$$

in which $C(K) = \Sigma C_m(K) = \Sigma C_m^{\underline{k}}(K)$ is the usual group of finite chains of K on coefficient field $\underline{k}$.

The presend section will show that the additive isomorphisms in the above form, as well as the known duality, can be realized by some "integration" procedure

$$\int : A^m(K) \to \mathrm{Hom}_{\underline{k}}(C_m(K), \underline{k})$$

to be defined below. The proof is again based on a modification of Weil's original one. First, we shall introduce a certain chain complex $W_* = \Sigma W_{m,p}$ somewhat dual to the Weil DGA W^* in the following way.

For each pair $m,p \geq 0$ let $W_{m,p}$ be the $\underline{k}$-module of collections $u = (u_J)$ where $u_J \in C_m(K_J)$ are zero, except only a finite number of them, however, with J running over all normally-ordered (p+1)-tuples of indices.

Dual to the diagram (D^W) of §III 3 we introduce now the following diagram

$$
(D_W)\qquad
\begin{array}{ccccccccccccc}
 & & \vdots & & \vdots & & & & \vdots & & \vdots & & \\
 & & \partial\downarrow & & \partial\downarrow\uparrow L & & & & \partial\downarrow\uparrow L & & \partial\downarrow\uparrow L & & \\
0 & \leftarrow & C_{p+1}(k) & \underset{P}{\overset{b}{\leftrightarrows}} & W_{0,p+1} & \underset{P}{\overset{b}{\leftrightarrows}} & \cdots & \leftrightarrows & W_{m,p+1} & \underset{P}{\overset{b}{\leftrightarrows}} & W_{m+1,p+1} & \leftrightarrows & \cdots \\
 & & \partial\downarrow & & \partial\downarrow\uparrow L & & & & \partial\downarrow\uparrow L & & \partial\downarrow\uparrow L & & \\
0 & \leftarrow & C_{p}(K) & \underset{P}{\overset{b}{\leftrightarrows}} & W_{0,p} & \underset{P}{\overset{b}{\leftrightarrows}} & \cdots & \leftrightarrows & W_{m,p} & \underset{P}{\overset{b}{\leftrightarrows}} & W_{m+1,p} & \leftrightarrows & \cdots \\
 & & \partial\downarrow & & \partial\downarrow\uparrow L & & & & \partial\downarrow\uparrow L & & \partial\downarrow\uparrow L & & \\
 & & \vdots & & \vdots & & & & \vdots & & \vdots & & \\
 & & \partial\downarrow & & \partial\downarrow\uparrow L & & & & \partial\downarrow\uparrow L & & \partial\downarrow\uparrow L & & \\
0 & \leftarrow & C_{0}(K) & \underset{P}{\overset{b}{\leftrightarrows}} & W_{0,0} & \underset{P}{\overset{b}{\leftrightarrows}} & \cdots & & W_{m,0} & \underset{P}{\overset{b}{\leftrightarrows}} & W_{m+1,0} & \leftrightarrows & \cdots \\
 & & \partial\downarrow & & \partial\downarrow\uparrow L & & & & \partial\downarrow\uparrow L & & \partial\downarrow\uparrow L & & \\
 & & 0 & \leftarrow & C_0(K) & \underset{b}{\leftarrow} & \cdots & \underset{b}{\leftarrow} & C_m(K) & \underset{b}{\leftarrow} & C_{m+1}(K) & \underset{b}{\leftarrow} & \cdots \\
 & & & & \downarrow & & & & \downarrow & & \downarrow & & \\
 & & & & 0 & & & & 0 & & 0 & &
\end{array}
$$

In the diagram (D_W)

$$\partial \text{ or } b : C_{r+1}(K) \to C_r(K)$$

both are the usual boundary operators in the chain complex $C(K)$. The other morphisms b, P, ∂, L occuring in the diagram are defined as follows;

<u>Definition</u> of $b : W_{0,p} \to C_p(K)$.

For

$$u = (u_J) \in W_{0,p}, \quad u_J \in C_0(K_J),$$

with J running over all normally-ordered (p+1)-tuples of indices, let $\mathrm{Ind}\, u_J$ be the sum of coefficients of the chain u_J. Then, by definition

$$bu = \Sigma\, \mathrm{Ind}\, u_J \cdot \sigma_J .$$

Remark that this is well-defined, since only a finite number of u_J can be non-zero and each u_J is a finite chain.

<u>Definition</u> of $b : W_{m+1,p} \to W_{m,p}$.

For $u = (u_J) \in W_{m+1,p}$ with $u_J \in C_{m+1}(K_J)$, J running over all normally ordered (p+1)-tuples of indices, let

$$b_J : C_{m+1}(K_J) \to C_m(K_J)$$

be the usual boundary operator. Then, by definition

$$bu = (b_J u_J) \in W_{m,p} .$$

Definition of P : $C_p(K) \to W_{0,p}$.

For $u = \Sigma u_J \sigma_J \in C_p(K)$ in which $u_J \in \underline{k}$ and Σ runs over normally-ordered $(p+1)$-tuples of indices $J = (j_0, \dots, j_p)$ we set by definition

$$P_J u = u_J v_{j_0} \in C_0(K_J) ,$$

$$Pu = (P_J u) \in W_{0,p} .$$

Definition of P : $W_{m,p} \to W_{m+1,p}$.

Let σ_J be any p-simplex of K with $J = (j_0, \dots, j_p)$ normally-ordered. For any normally-ordered m-simplex $\tau = (v_{k_0} \dots v_{k_m}) \in K_J$, let $P_J \tau = 0$ if some $k_j = j_0$ while

$$P_J \tau = (-1)^{\nu} \cdot (v_{k_0} \dots v_{k_{\nu-1}} v_{j_0} v_{k_\nu} \dots v_{k_m})$$

if no $k_i = j_0$ and $k_{\nu-1} < j_0 < k_\nu$. Then, for

$$u = (u_J) \in W_{m,p} , \quad u_J = \Sigma u_{J\tau} \tau \in C_m(K_J) , \quad u_{J,\tau} \in \underline{k} ,$$

we set by definition

$$Pu = (P_J u_J) \in W_{m+1,p} ,$$

$$P_J u_J = \Sigma u_{J,\tau} P_J \tau \in C_{m+1}(K_J) .$$

Definition of ∂ : $W_{m,0} \to C_m(K)$.

For $x = (x_j) \in W_{m,0}$, $x_j \in C_m(K_j)$, we set by definition, as is clearly legitimate,

$$\partial x = \Sigma x_j \in C_m(K) .$$

Definition of ∂ : $W_{m,p+1} \to W_{m,p}$.

Let $x = (x_H) \in W_{m,p+1}$, $x_H = \Sigma x_{H,\tau} \tau \in C_m(K_H)$, with $x_{H,\tau} \in \underline{k}$ and H running over all normally ordered $(p+2)$-tuples of indices. Consider any normally ordered $(p+1)$-tuple of indices $J = (j_0, \dots, j_p)$. For any vertex $v_k \in K_J$ with $k \notin J$ we shall set by definition

$$\partial x = (y_J) \in W_{m,p} ,$$

$$y_J = \Sigma\, \varepsilon_{kJ}\, x_{[kJ],\tau}\, \tau \in C_m(K_J) ,$$

in which $[kJ]$ means the normally-ordered (p+2)-tuple formed from $(k, j_0, \ldots, j_p)$ and $\varepsilon_{kJ} = +1$ or -1 according to the permutation to bring $(k, j_0, \ldots, j_p)$ into the normal order is even or odd.

<u>Definition</u> of $L: C_m(K) \to W_{m,0}$.

For any normally-ordered m-simplex τ of K let the first vertex of τ be $v_{i(\tau)}$. For $z = \Sigma\, z_\tau \tau \in C_m(K)$ with $z_\tau \in \underline{k}$ we set now by definition

$$Lz = (x_j) \in W_{m,0} ,$$

$$x_j = \sum_{i(\tau)=j} z_\tau \tau \in C_m(K_j).$$

<u>Definition</u> of $L: W_{m,p} \to W_{m,p+1}$.

Let $x = (x_J) \in W_{m,p}$, $x_J = \Sigma\, x_{J,\tau} \tau \in C_m(K_J)$ with $x_{J,\tau} \in \underline{k}$ and J running over all normally-ordered (p+1)-tuples of indices. For any normally-ordered (p+2)-tuple of indices $H = (h_0, \ldots, h_{p+1})$ let $H_\nu = (h_0, \ldots, \hat{h}_\nu, \ldots, h_{p+1})$. Since $K_H \subset K_{H_\nu}$ we may put

$$y_{H,\tau} = (-1)^\nu \cdot x_{H_\nu,\tau}$$

if the first vertex of τ is some v_{h_ν} and

$$y_{H,\tau} = 0$$

otherwise. We define then

$$Lx = (y_H) \in W_{m,p+1} ,$$

$$y_H = \Sigma\, y_{H,\tau}\, \tau \in W_{m,p+1} .$$

About the morphisms in the diagram we have the following lemmas. All the proofs are only direct verifications and will be omitted throughout.

<u>Lemma 1</u>: For $m,p \geq 0$,

$$b^2 = 0 : W_{m+2,p} \to W_{m,p} ,$$

and

$$\partial^2 = 0 : W_{m,p+2} \to W_{m,p} .$$

Lemma 2: For the morphisms P, b we have

$$Pb + bP = \text{ident.} \mid W_{m,p} ,$$

$$bP = \text{ident.} \mid C_p(K) .$$

Lemma 3: For the morphisms L, ∂ we have

$$L\partial + \partial L = \text{ident.} \mid W_{m,p} ,$$

$$\partial L = \text{ident.} \mid C_m(K) .$$

Lemma 4: The part of the diagram (D_W) involving only b and ∂ is commutative.

Lemma 5: In the diagram (D_W) the sequences

$$0 \longleftarrow C_p(K) \underset{b}{\longleftarrow} W_{0,p} \underset{b}{\longleftarrow} \cdots \underset{b}{\longleftarrow} W_{m,p} \underset{b}{\longleftarrow} \cdots$$

of the horizontal rows are exact. The same is true for the sequences

$$\cdots \underset{\partial}{\longrightarrow} W_{m,p} \underset{\partial}{\longrightarrow} \cdots \longrightarrow W_{m,0} \underset{\partial}{\longrightarrow} C_m(K) \longrightarrow 0$$

of the vertical columns.

Setting now

$$W_* = \Sigma W_m ,$$

$$W_m = \Sigma W_{q,m-q} ,$$

$$\bar{\partial} = (-1)^m \partial \mid W_{m,p} ,$$

$$D = b + \bar{\partial} .$$

In view of the above Lemmas we see that W_* becomes a chain complex with boundary operator D.

Definition: The chain complex $W_* = \Sigma W_m$ with D as boundary operator will be called the Weil Chain Complex associated to the complex K.

The following lemmas are also easily proved by induction in using the above lemmas:

Lemma 6: Let $w \in A^m(K)$ be a cycle in $A^*(K)$: $dw = 0$. Then, $\delta(I\delta)^q w \in W^{m-q,g}(K)$ is also a cycle in Weil DGA W^* for any $q \geq 0$. In particular

$$d\delta(I\delta)^q w = 0 \; .$$

Lemma 7: Let $x \in C_m(K)$ be a cycle in $C(K)$: $\partial x = bx = 0$. Then $(\partial P)^q x \in W_{q-1,m-q}$ is also a cycle in the Weil Chain complex W_* for any $q \geq 1$. In particular,

$$b(\partial P)^q x = 0 \; .$$

As in §III 3, we now have the following

Theorem 1: $(\partial P)^{m+1} : C_m(K) \to C_m(K)$ is an isomorphism which is at most a sign change:

$$(\partial P)^{m+1} = (-1)^{\frac{1}{2}m(m+1)} : C_m(K) \approx C_m(K) \; .$$

Proof: It is sufficient to consider an elementary chain $z = \sigma_J \in C_m(K)$ where $J = (j_0, \ldots, j_m)$ is a normally-ordered (m+1)-tuple of indices. Set $J_i' = (j_0, \ldots, j_i)$, $J_i'' = (j_i, \ldots, j_m)$, $J_{i,\nu}'' = (j_i, \ldots, \hat{j}_\nu, \ldots, j_m)$ for $0 \leq i \leq m$ and $i \leq \nu \leq m$. Let

$$P(\partial P)^i z = (x_{H_i}^{(i)}) = x^{(i)} \in W_{i,m-i} , \qquad \text{for } 0 \leq i \leq m ,$$

$$(\partial P)^{i+1} z = (y_{L_i}^{(i)}) = y^{(i)} \in W_{i,m-i-1} , \qquad \text{for } 0 \leq i \leq m-1 ,$$

$$(\partial P)^{m+1} z = y^{(m)} \in C_m(K) ,$$

in which H_i and L_i run over all normally-ordered (m-i+1)- and (m-i)-tuples of indices respectively.

By induction we verify now directly:

For $0 \leq i \leq m$,

$$x_{H_i}^{(i)} = \begin{cases} (-1)^{\frac{1}{2}i(i+1)} \cdot \sigma_{J_i'} \in C_i(K_{J_i''}) , & \text{for } H_i = J_i'' , \\ 0 , & \text{for } H_i \neq J_i'' . \end{cases}$$

For $0 \leq i < m-1$,

$$y_{L_i}^{(i)} = \begin{cases} (-1)^{\nu-i} \cdot (-1)^{\frac{1}{2}i(i+1)} \cdot \sigma_{J_i'} \in C_i(K_{J_{i,\nu}''}) , & \text{for } L_i = J_{i,\nu}'' , \\ 0 , & \text{for } L_i \neq \text{any } J_{i,\nu}'' . \end{cases}$$

Whence, we get

$$(\partial P)^{m+1} z = \partial x^{(m)} = \sum_{j \in N} x_j^{(m)} = x_{j_m}^{(m)}$$
$$= (-1)^{\frac{1}{2}m(m+1)} \cdot \sigma_{J_m'} = (-1)^{\frac{1}{2}m(m+1)} \cdot z$$

The $\underline{k}$-module $W_{m,p}$ will now be made dual to $W^{m,p}$ by means of an integration or scalar product to be defined as follows.

First let us set by definition for any normally-ordered m-simplex σ_H of K with $H = (h_0, \dots, h_m)$,

$$\int_{\sigma_H} t_{h_0}^{s_0} \dots t_{h_m}^{s_m} dt_{h_0} \dots \widehat{dt_{h_\nu}} \dots dt_{h_m}$$
$$= (t_{h_0}^{s_0} \dots t_{h_m}^{s_m} dt_{h_0} \dots \widehat{dt_{h_\nu}} \dots dt_{h_m}, \sigma_H)$$
$$= (-1)^\nu \cdot \frac{s_0! \dots s_m!}{(\Sigma s_i + m)!} .$$

For any $\alpha \in A^m(\sigma_H)$, we then extend the definition of $\int_{\sigma_H} \alpha$ or (α, σ_H) by linearity. This definition is legitimate since it is readily verified that $(\alpha, \sigma_H) = 0$, whenever α is in the ideal generated by $t_{h_0} + \dots + t_{h_m} - 1$ and $dt_{h_0} + \dots + dt_{h_m}$.

<u>Definition:</u> For any form

$$w = (w_J) \in W^{m,p}$$

and any chain

$$x = (x_J) \in W_{m,p}$$

in which J runs over all normally-ordered (p+1)-tuples of indices, the <u>scalar product</u> between w and x is the value (w,x) given by

$$(w,x) = \sum_J (w_J, x_J) ,$$

in which

$$(w_J, x_J) = \sum_\tau x_{J,\tau} (w_J(\tau), \tau) ,$$
$$x_J = \sum_\tau x_{J,\tau} \tau \in C_m(K_J), \quad x_{J,\tau} \in \underline{k} ,$$

Note that the definition is legitimate since Σ_J has only a finite number of non-zero terms and each x_J is a finite chain.

<u>Definition:</u> For any elements

$$w \in W^{m,p}, \quad x \in W_{m,p},$$

the value $(w,x) \in \underline{k}$ will be called the <u>integral</u> of w over x to be denoted by

$$\int_x w \text{ or } (w,x) .$$

The module-morphism

$$\int : A^m(K) \to \mathrm{Hom}_{\underline{k}}(C_m(K), \underline{k})$$

given by

$$(\int w)(x) = \Sigma\, x_J \int_{\sigma_J} w(\sigma_J)$$

for $w \in A^m(K)$, $x = \Sigma\, x_J \sigma_J \in C_m(K)$, $x_J \in \underline{k}$, and J runs over normally-ordered $(m+1)$-tuples of indices, will be called the <u>integration</u> of $A^m(K)$ on K, and $(\int w)(x)$ will also be denoted simply as $\int_x w$ or (w,x), or also $w(x)$ if no misunderstanding can occur.

<u>Theorem 2</u>: For any given elements

$$w \in W^{m,p}, \quad x \in W_{m+1,p}, \quad y \in W_{m,p+1},$$
$$u \in C^p(K), \quad \xi \in A^m(K),$$
$$\alpha \in W_{0,p}, \quad z \in W_{m,0},$$

we have

$$\int_x dw = \int_{bx} w, \quad \text{or} \quad (dw,x) = (w,bx).$$
$$\int_y \delta w = \int_{\partial y} w, \quad \text{or} \quad (\delta w,y) = (w,\partial y).$$
$$\int_\alpha du = \int_{b\alpha} u, \quad \text{or} \quad (du,\alpha) = (u,b\alpha).$$
$$\int_z \delta\xi = \int_{\partial z} \xi, \quad \text{or} \quad (\delta\xi,z) = (\xi,\partial z).$$

<u>Proof</u>: By direct verifications, which will be omitted.

We now come to the main result of this section, viz.

<u>Theorem 3</u>: The integration

$$\int : A^*(K) \to \mathrm{Hom}_{\underline{k}}(C(K),\underline{k}) \approx C^*(K)$$

commutes with differentiations in $A^*(K)$ and $\mathrm{Hom}_{\underline{k}}(C(K),\underline{k})$ and induces additive isomorphisms

$$\int_H : H_m(A^*(K)) \approx H_m(C^*(K))$$

which realize the isomorphism in the deRham-Sullivan Theorem.

Proof: Using Theorem 1 and then Theorem 2 we have for $w \in A^m(K)$, $x \in C_m(K)$ with $dw = 0$ and $bx = 0$,

$$(-1)^{\frac{1}{2}m(m+1)} \cdot (w,x) = (w,(\partial P)^{m+1}x)$$

$$= (\delta w, P(\partial P)^m x) = ((Id + dI)\delta w, P(\partial P)^m x)$$

$$= (dI\delta w, P(\partial P)^m x)$$

$$= (I\delta w, bP(\partial P)^m x)$$

$$= (I\delta w, (1-Pb)(\partial P)^m x)$$

$$= (I\delta w, (\partial P)^m x), \quad \text{by Lemma 7}.$$

Proceeding inducitively in this manner we finally get

$$(-1)^{\frac{1}{2}m(m+1)} \cdot (w,x) = ((I\delta)^{m+1} w)(x) .$$

By Theorem of the last section,

$$d'\delta = (-1)^{\frac{1}{2}m(m+1)} \cdot (I\delta)^{m+1} : A^m(K) \to C^m(K)$$

induces an isomorphism

$$(d'\delta)_H : H(A^*(K)) \simeq H(C^*(K))$$

and thus, we see that isomorphism is realized by the integration $\int_x w$, as to be proved.

The following theorem is quite evident:

Theorem 3: For any simplicial map f of a connected countable simplicial complex K into another such one L, there will induce a natural DGA-morphism

$$f^A : A^*(L) \to A^*(K)$$

and a natural module-morphism

$$f_C : C(K) \to C(L) .$$

For the integration we then have

$$\int_x f^A w = \int_{f_C x} w$$

for any $w \in A^m(L)$ and $x \in C_m(K)$.

Corollary: The ring-morphism

$$(f^A)_H : H(A^*(L)) \to H(A^*(K))$$

induced by f^A coincides with the usual morphism

$$f^H : H^*(L) \to H^*(K)$$

induced by f via deRham-Sullivan Theorem.

III.5 HOMOTOPY INVARIANCE AND CALCULABILITY OF I^*-MEASURE

For the category of arbitrary connected countable simplicial complexes in weak topology K we have associated a DGA-measure $I^*(K) = I^*_{\underline{k}}(K)$ for each field $\underline{k}$ of characteristic 0. In this section, we shall now investigate the adequacy of such a measure from the point of view indicated in §I 3.

For the finiteness condition we see from Chap. II, that the I^*-measure of a complex is usually finitely generated in each degree, e.g., in the case of simply-connected finite complexes. The requirement of finiteness is thus observed in the main, at least in case of complexes not too complicated.

For the invariance condition let us first prove the following

Theorem 1: The I^*-measure of connected countable simplicial complexes in weak topology is combinatorically invariant.

Proof: Let K' be a simplicial subdivision of such a simplicial complex K. Consider any simplicial map s of K' to K which sends any vertex v' of K' to one of K which is a vertex of a simplex K containing v' in its interior. Then, s will induce a DGA-morphism

$$s^A : A^*(K) \to A^*(K')$$

which will induce in turn a ring-morphism

$$(s^A)_H : H(A^*(K)) \to H(A^*(K')) .$$

By the deRham-Sullivan Theorem this is the same as the morphism

$$s^H : H^*(K) \to H^*(K')$$

which is known to be an algebraic isomorphism. Thus, s^A is an H-isomorphism, so that by §II 4., we have

$$\min A^*(K) \simeq \min A^*(K')$$

or

$$I^*(K) \simeq I^*(K') .$$

this proves the combinatorical invariance of $I^*(K)$.

More generally, we have the following

<u>Theorem 2</u>: The I^*-measure of connected countable simplicial complexes with weak topology is homotopically invariant.

<u>Proof</u>: Let K,L be such simplicial complexes, which are homotopically equivalent with the continuous map $g : K \to L$, realizing their homotopical equivalence. Let K' be a certain simplicial subdivision of K and $s' : K' \to L$ a simplicial map, which is a simplicial approximation of g. Let $s : K' \to K$ be a simplical map, as in the proof of Theorem 1. Then, we have a diagram of simplicial maps

$$K \xleftarrow{s} K' \xrightarrow{s'} L$$

which will induce a diagram of DGA-morphisms

$$A^*(K) \xrightarrow{s^A} A^*(K') \xleftarrow{s'^A} A^*(L) .$$

This, in turn, will induce a diagram of ring-morphisms

$$H(A^*(K)) \xrightarrow{(s^A)_H} H(A^*(K')) \xleftarrow{(s'^A)_H} H(A^*(L))$$

which, by Theorem of deRham-Sullivan, is the same as the diagram below:

$$H^*(K) \xrightarrow{s^H} H^*(K') \xleftarrow{s'^H} H^*(L) .$$

Now, it is known that both s^H and s'^H are ring-isomorphisms, or that both s^A and s'^A are H-isomorphisms. Therefore, it follows again from §II 4. that

$$I^*(K) \simeq I^*(K') \simeq I^*(L)$$

or I^*-measure is homotopically-invariant, as to be proved.

The above theorem shows, that the following definition is legitimate:

Definition: For a connected space X in the HSC-category, i.e., one homotopically equivalent to connected countable simplicial complexes in weak topology, the I^*-measure of any such simplicial complex will be called the I^*-measure of the space X and will be denoted by $I^*(X)$.

Consider two spaces X, Y in the HSC-category and a continuous map $f: X \to Y$. Let $g: K \to X$, $h: Y \to L$ be some homotopic equivalence with connected countable simplicial complexes K, L. Let K' be a simplicial subdivision of K and $f': K' \to L$ be a simplicial approximation of $hfg: K \to L$. For the induced DGA-morphism

$$f'^A : A^*(L) \to A^*(K')$$

there will be associated some minimal morphisms

$$\min f'^A : \min A^*(L) \to \min A^*(K')$$

or

$$\min f'^A : I^*(Y) \to I^*(X) .$$

By Chap. II it is clear that $\min f'^A$ is uniquely determined up to homotopy, and any such morphism will be said to be induced by f and denoted as

$$f^I : I^*(Y) \to I^*(X) .$$

Finally, concerning the calculability condition let us first remark, that the I^*-measure may be non-calculable with respect to quite simple geometrical constructions, and the following is such an example.

Example: Let K be the projective plane and L a projective line in K with inclusion $i: L \subset K$.
Then, we have

$$I^*(K) = 0$$

$$I^*(L) = \mathrm{Extr}\,(x)$$

with $\deg x = 1$ and the DGA-morphism induced by i is the trivial one

$$i^I : \begin{array}{ccc} I^*(K) & \longrightarrow & I^*(L) \\ \| & & \| \\ 0 & \longrightarrow & \mathrm{Extr}\,(x) \end{array}$$

Now, let K be the ordinary 3-space R^3 and L be any knot in R^3, with $i: L \subset K$

again the inclusion. Then, $I^*(K)$, $I^*(L)$ and the induced DGA-morphism i^I will be same as above.

Remove now L from K to get the space $K-L$. Then, it is clear that in the first case $K-L$ is contractible, so that $I^*(K-L)=0$, while in the second case $I^*(K-L)$ is non-trivial. This shows that the I^*-measure is not calculable with respect to the geometrical construction in removing a subspace from a space.

In spite of the non-calculability example above, we shall show in later chapters that the I^*-measure is calculable with respect to most of the important geometrical constructions, usually occuring in algebraic topology. This is in fact the main task to which the present book is devoted.

Let us add some final remarks.

(A) The notion of deRham-Sullivan algebra and hence also the I^*-measure can be easily extended to much broader categories of spaces via, e.g., the singular complex or nerve complex of coverings, which we shall not enter.

(B) Any field $\underline{k}$ of characteristic 0 contains the rational field $\underline{Q}$ as a subfield so that, as easily seen,

$$I^*_{\underline{k}}(X) \approx I^*_{\underline{Q}}(X) \otimes \underline{k}$$

in a natural manner.

(C) In defining $A^*(K)$ we have used forms in barycentric coordinates on each simplex with polynomial coefficients. Forms with C^∞- or other kinds of functions as coefficients may be used equally well in getting a theorem of deRham-Sullivan type, with due modifications of the proof given in this chapter. The minimal model thus deduced will then be simply $I^*_{\underline{R}}(K)$, giving nothing in new.

(D) Any (connected) differential manifold M, according to J.H.C. Whitehead, possesses a canonical class of simplicial subdivisions combinatorically equivalent to each other. Let K be any such associated complex. Then, the usual DGA-algebra $A^*(M)$ of differential forms with, e.g., C^∞-functions as coefficients on M, will induce canonically a subalgebra of the deRham-Sullivan algebra $A^*(K)$ of K, defined in using the same kind of coefficients as in (C).
The induced inclusion morphism is an H-isomorphism, as a consequence of the deRham-Sullivan theorem. It follows that the usual $A^*(M)$ can be used for the determination of $I^*_{\underline{R}}(M)$, viz. $I^*_{\underline{R}}(M) \approx \min A^*(M)$.

(E) For a space X in the HCS-category the cohomology-ring of X on any coefficients field $\underline{k}$ of characteristic 0 is completely determined by its I^*-measure, viz,

$$H^*_{\underline{k}}(X) \approx H(I^*_{\underline{k}}(X)) .$$

The converse is surely not true in general.

(F) In certain cases $I^*(X)$ of a space X can, however, be determined from the knowledge of its cohomology $H^*(X)$ alone, viz. $I^*(X) \approx \min H^*(X)$, in considering $H^*(X)$ as a DGA with a trivial differential. Such a space will then be said to be formal. We list a few of such spaces with the corresponding I^*-measure, below. These are all easily determined by means of methods developed in this chapter and the preceding one.

(a) Sphere S^n of dimension n :

$$I^*(S^n) \approx \text{Extr}\,(x) ,$$

with $\deg x = n$, for n odd.

or

$$I^*(S^n) \approx \text{Polym}\,(y) \otimes \text{Extr}\,(x) ,$$

with

$$\deg x = 2n - 1 , \qquad \deg y = n ,$$
$$dx = y^2 , \qquad dy = 0$$

for n even.

(b) Complex projective spaces CP_n of complex dimension n :

$$I^*(CP_n) \approx \text{Polym}\,(y) \otimes \text{Extr}\,(x)$$

with

$$\deg x = 2n + 1 , \qquad \deg y = 2 ,$$
$$dx = y^{n+1} , \qquad dy = 0 .$$

(c) I^*-measure of the Eilenberg-MacLane space $(K(\underline{Q},n)$ is given by $I^*(K(\underline{Q},n)) \approx H^*(K(\underline{Q},n)) \approx \text{Free}\,(x)$, which is, according to n being odd or even, either an exterior algebra or a polynomial algebra on a single generator x of degree n .

(d) The union of two spheres at a single point $S^m \vee S^n$ with

$$I^*(S^m \vee S^n) \approx \min\,(I^*(S^m) \oplus I^*(S^n)) \approx \min\,(H^*(S^m) \oplus H^*(S^n))$$

which may be described in full as in the known theorem of Hilton-Milnor. Similarly, for any bouquet of spheres.

(e) Riemannian symmetric spaces and more generally certain homogeneous spaces G/H with G a compact connected Lie group and H a closed connected group, such that (G,H) form a Cartan pair, which we shall not explicit here, cf. e.g. Chap. V.

(f) Compact Kähler manifolds, cf. Déligne, et al [1].

Chapter IV

I^* - MEASURE AND HOMOTOPY

IV.1 THE CARTAN-SERRE EXTENSION OF A DGA AND THE HUREWICZ NUMBER

Definition: A DGA $\mathcal{A}$ will be said to be of type n $(n > 1)$ if as a GA, it is free of the form

$$(1) \qquad \mathcal{A} = \mathrm{Free}\,(\bar{z}_i, z_i) \otimes \mathrm{Free}\,(x_j)$$

with i,j each running over some index sets and with degree and differential verifying the following conditions:

$$(2) \qquad \begin{cases} \deg \bar{z}_i + 1 = \deg z_i < n\,, \\ \mathrm{Min}\deg x_j = n\,, \\ d\bar{z}_i = z_i\,, \\ dx_j \in \mathcal{A}^+ \cdot \mathcal{A}^+\,, \\ dx_j \text{ contains no } x_k \text{ of degree} \geq \deg x_j\,. \end{cases}$$

The last condition means that no $\bar{z}_i$ of degree 1 will appear in dx_j.

Lemma: For a DGA $\mathcal{A}$ of type n in the form (1) verifying (2) we can replace the free generators x_j by x'_j of the form

$$(3) \qquad x'_j = x_j - Q_j(\bar{z}_i, z_i, x'_k)$$

with Q_j certain polynomials in $\bar{z}_i, z_i, x'_k$ such that the following conditions are verified:

$$(4) \qquad \begin{cases} Q_j \text{ contains only } x'_k \text{ of degree} < \deg x_j\,, \\ \mathcal{A} \approx \mathrm{Free}\,(\bar{z}_i, z_i) \otimes \mathcal{B}\,, \\ \mathcal{B} = \mathrm{Free}\,(x'_j)\,, \\ dx'_j = 0 \text{ if } \deg x'_j = n\,, \\ dx'_j \in \mathcal{B}^+ \cdot \mathcal{B}^+. \end{cases}$$

Proof: First, let us consider any generator x_j of degree n.

The degree conditions imply that

$$dx_j = P_j(\bar{z}_i, z_i)$$

for some polynomial P_j in $\bar{z}_i, z_i$. Then, $dP_j = 0$ would imply

$$P_j = dQ_j(\bar{z}_i, z_i)$$

for some polynomial Q_j in $\bar{z}_i, z_i$. Set

$$x'_j = x_j - Q_j(\bar{z}_i, z_i).$$

Then

$$dx'_j = 0$$

so that the assertion of the lemma is verified for the generators of degree n.

Suppose it has been proved that for any x_j of degree $\leq n+m$ ($m \geq 0$ fixed) it has been replaced by some

$$x'_j = x_j - Q_j(\bar{z}_i, z_i, x'_k)$$

with some polynomial Q_j such that Q_j contains besides $\bar{z}_i, z_i$ only x'_k of degree $< \deg x_j$, while dx'_j contains no more any $\bar{z}_i$ or z_i. Then we shall prove that this can also be done for any generator x_j of degree $n+m+1$. This induction procedure will prove the truth of the lemma and give the expressions (3), (4), as required.

For this purpose let us write for simplicity

$$X'^N_J = x'^{n_1}_{j_1} x'^{n_2}_{j_2} \cdots x'^{n_s}_{j_s}$$

for power products in x' with J standing for the index set $(j_1, \ldots, j_s)$, and N the exponent set $(n_1, \ldots, n_s)$. Then, for any x_j of degree $n+m+1$ we can, by induction, write dx_j in the form

$$dx_j = \sum X'^N_J \cdot P^N_J(\bar{z}_i, z_i)$$

with P^N_J some polynomials in $\bar{z}_i, z_i$ alone and J, N running over some index sets. The minimum of the degree of power products X'^N_J occuring in the above expression of dx_j will be called provisionally the <u>rank</u> of x_j, to be denoted by Rank x_j. Differentiating again dx_j we will get an expression of the form

$$0 = \sum_{(J,N)} dX'^N_J \cdot P^N_J + \sum_{(J,N)} \varepsilon^N_J X'^N_J \cdot dP^N_J ,$$

in which $\varepsilon_J^N = +1$ or -1 according to $\deg X'^N_J$ = even or odd. Again by induction, any dx'_k with x'_k occuring in some X'^N_J will contain no $\bar{z}_i$ or z_i so that any power product of x' occuring in any dX'^N_J will have a degree greater by 1 than that of X'^N_J. It follows that, if J^0, N^0 are any of the index sets appearing in the summation for which the degree of corresponding $X'^{N^0}_{J^0}$ is a minimum, then the corresponding term cannot be cancelled by the others and we necessarily have $dP^{N^0}_{J^0} = 0$.

As $P^{N^0}_{J^0} \in \text{Free}(\bar{z}_i, z_i)$ so

$$P^{N^0}_{J^0} = dQ^{N^0}_{J^0}$$

for some polynomial $Q^{N^0}_{J^0}$ in $\bar{z}_i, z_i$ alone.

Now, let us replace x_j by some

$$x^*_j = x_j - \Sigma^0 \varepsilon^{N^0}_{J^0} X'^{N^0}_{J^0} Q^{N^0}_{J^0},$$

in which Σ^0 is the summation over index sets (J^0, N^0) for which $\deg X'^{N^0}_{J^0}$ attains the minimum value. Then

$$dx^*_j = \Sigma^* X'^{N^*}_{J^*} P^{N^*}_{J^*} - \Sigma^0 \varepsilon^{N^0}_{J^0} dX'^{N^0}_{J^0} Q^{N^0}_{J^0},$$

in which Σ^* is the summation over index sets (J^*, N^*) which is just the index set of (J, N) with all (J^0, N^0) removed.

It is clear that $\text{Rank}\, x^*_j > \text{Rank}\, x_j$. Hence, on repeating the process we will finally replace x_j by some one x'_j for which the rank can no more be raised. In other words, in dx'_j no more polynomials P^N_J in $\bar{z}_i, z_i$ only can occur. This completes the induction and thus, the lemma is proved.

Definition: A DGA $\mathcal{A}$ of type n in the form (1), (2) with conditions of the lemma verified, will be said to be in normal form.

The lemma says that any such DGA of type n can always be put in normal form or normalized by choosing suitable free generators.

Let $\mathcal{A}$ be now a DGA of type n already in normal form, as given in (4). Then, it is clear that

$$\min \mathcal{A} \approx \min \mathcal{B} \approx \mathcal{B}.$$

Let us separate the generators (x'_j) into two parts: (x'_k) of degree n and (x'_l) of degree $> n$ with k, l running over respective index sets.

Corresponding to each x'_k of degree n let us adjoin a new generator x'^-_k to A to form a new DGA

$$\tilde{A} = A \otimes \text{Free}(x'^-_k)$$

with degree and differential

$$\deg x'^-_k = n-1 , \quad dx'^-_k = x'_k .$$

Then, as a GA,

$$\tilde{A} = \text{Free}(z^-_i , z_i , x'^-_k , x_k) \otimes \tilde{B}$$

with

$$\tilde{B} = \text{Free}(x'_l)$$

clearly is a DGA of type $n+m$ where $n+m\ (>n)$ is the minimum of the degrees of x'_l. We shall now lay down the following

Definition: The above extended DGA $\tilde{A}$ will be called the Cartan-Serre extension of A.

Definition: For any DGA A the number of free generators of degree s, if it is finite, of $\min A$, will be called the Hurewicz number in degree s of A, and will be denoted by $q_s(A)$.

For any DGA A of type n given by (1), (2) and normalized to $A \approx \text{Free}(z^-_i , z_i) \otimes B$, as given in (4), we see from the proof of the lemma that the numbers of free generators in each degree $\geq n$ are the same for A and $\min A \approx B$. It follows that q_s = number of free generators of degree s in A for $s > n$. From this remark we get the following

Proposition: Let A be a DGA of type n and $\tilde{A}$ its Cartan-Serre extension, then for $s > n$,

$$q_s(\tilde{A}) = q_s(A) .$$

IV.2 THE CARTAN-SERRE EXTENSION OF A SPACE

Definition: A space X will be said to be of finite type if all homotopy and homology group-measures are finitely generated. The rank of the s-th homotopy group of such a space is then well-defined and will be called the s-th Hurewicz number of the space and will be denoted by

$$q_s(X) = \text{Rank}\ \pi_s(X) = \text{Dim}_{\underline{Q}}\ \pi_s(X) \otimes \underline{Q}\ .$$

It will be said to be of type n $(n > 1)$ if

$$\pi_s(X) \otimes \underline{Q} = 0\ , \quad \text{for} \quad s < n\ , \quad \text{while} \quad \pi_n(X) \otimes Q \neq 0\ ,$$

or

$$q_s(X) = 0 \qquad \text{for} \quad s < n\ , \quad \text{while} \quad q_n(X) \neq 0\ .$$

Let X be now a connected and simply connected space of finite type having the same homotopy type as a simplicial complex. For simplicity let us denote the homotopy groups of X by

$$(1) \qquad \pi_s = \pi_s(X)\ .$$

Suppose that X is of type n for some $n > 1$ so that

$$(2) \qquad \pi_i = \text{finite}\ , \qquad i \leq n-1$$

while

$$(3) \qquad \pi_n \approx \underbrace{\underline{Z} \oplus \ldots \oplus \underline{Z}}_{r} \oplus \text{ a finite group } F_n \approx r\underline{Z} \oplus F_n\ , \quad \text{or}$$

$$(3)' \qquad \pi_n \otimes \underline{Q} \approx \underbrace{\underline{Q} \oplus \ldots \oplus \underline{Q}}_{r} \approx r\underline{Q}\ .$$

Then, by the generalized Hurewicz Theorem of Serre's abelian class theory, we also have

$$(4) \qquad H_s^{\underline{Q}}(X) = 0\ , \qquad 0 \leq s \leq n-1\ ,$$

while

$$(5) \qquad H_n^{\underline{Q}}(X) \approx \pi_n \otimes \underline{Q} \approx r\underline{Q}\ .$$

In particular, if we denote by $p_n(X)$ the n-th betti number of X, then

$$(6) \qquad \text{Rank}\ \pi_n = \dim_{\underline{Q}} H_n^{\underline{Q}}(X) = \dim_{\underline{Q}} H_{\underline{Q}}^n(X) = p_n(X)\ (=r)\ .$$

Moreover, $H_n^{\underline{Q}}(X)$ is the natural image of $H_n^{\underline{Z}}(X)$.

Suppose that the space X is of type $n > 1$ as above with $r \geq 1$. Let $K_{n,r}$ be an Eilenberg–MacLane space

$$K_{n,r} = K(\underbrace{\underline{Q} \otimes \ldots \otimes \underline{Q}}_{r},n) = \underbrace{K(\underline{Q},n) \times \ldots \times K(\underline{Q},n)}_{r} .$$

Consider the universal fibration

$$(\mathcal{K}) \qquad K_{n-1,r} \subset P_{n,r} \xrightarrow{g'} K_{n,r}$$

with fiber $K_{n-1,r}$ also being an Eilenberg-MacLane space $\underbrace{K(\underline{Q},n-1) \times \ldots \times K(\underline{Q},n-1)}_{r}$, projection g' and a contractible space $P_{n,r}$ as the fiber space.

We shall denote the fundamental classes in $H^n_{\underline{Q}}(K(\underline{Q},n))$ of the j-th copy $K(\underline{Q},n)$ in $K_{n,r}$ by Ξ_j.

Now, take a basis X_j of $H^n_{\underline{Q}}(X)$ and define a map

$$f : X \to K_{n,r}$$

$$f^H \Xi_j = X_j , \qquad j = 1, \ldots, r .$$

Then, under f there will be an induced fibration

$$(\mathcal{F}) \qquad K_{n-1,r} \subset \tilde{X} \xrightarrow{g} X$$

from $(\mathcal{K})$ such that there is a commutative diagram of maps below:

(D_1)

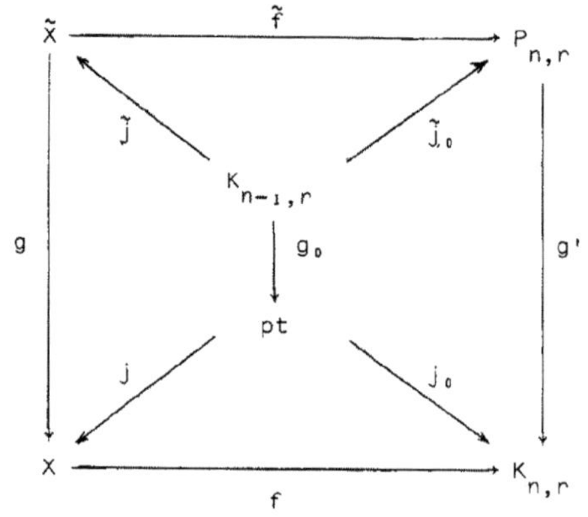

Now, according to § 1.4 we can realize the spaces in the fibration $(\mathcal{K})$ as simplicial complexes, with inclusion $\tilde{j}_0$ and projection g' both being simplicial maps. In replacing X by a simplicial complex of the same homotopy type and f

by a simplicial approximation we shall assume, that all spaces in the above diagram are simplicial complexes, and all maps are also simplicial ones. There is no loss of generality in assuming this, since we are now interested in the case of a homotopic character only.

From the homotopy exact sequence connected with a fibration, we have now

$$\pi_s(\tilde{X}) \otimes \underline{Q} = 0 , \qquad 1 \leq s \leq n ,$$

$$\pi_s(\tilde{X}) \otimes \underline{Q} \approx \pi_s(X) \otimes \underline{Q} = \pi_s \otimes \underline{Q} , \qquad s > n .$$

Consequently, $\tilde{X}$ will also be a space of type $\geq n+1$.

<u>Definition</u>: The space $\tilde{X}$ constructed as above from a space X of type n will be called the <u>Cartan-Serre extension</u> of the space X.

<u>Remark</u>: In case $\pi_n(X) \otimes \underline{Q} = 0$, so that X actually is of type $n+m$ for some $m \geq 1$, then the Cartan-Serre extension $\tilde{X}$ will be taken to be coincident with X.

It is clear that for the space X of type n the I^*-measure $I^*(X)$ of X is a DGA of type n. We have now the following

<u>Proposition</u>: Let X be a space of type n for which the I^*-measure is the minimal model of a DGA $\mathcal{A}$ of type n. Then, the I^*-measure of the Cartan-Serre extension $\tilde{X}$ of X is the minimal model of the Cartan-Serre extension $\tilde{\mathcal{A}}$ of $\mathcal{A}$.

<u>Proof</u>: Since all the spaces and maps involved in (D_1) are simplicial we shall have a commutative diagram of the following form:

(D_2)

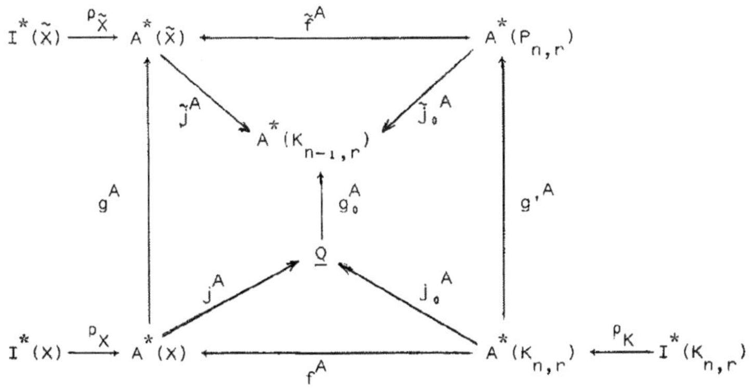

In the diagram ρ_K, ρ_X both are associated minimal morphisms. For the I^*-measure we may take

$$I^*(K_{n,r}) \approx \mathrm{Free}\,(\xi_j)\,,$$
$$I^*(X) \approx \mathrm{Free}\,(x_j\,,\,x_k)$$

in which (k running over some index set)

$$\deg \xi_j = n\,, \qquad \deg x_j = n\,,$$
$$\deg x_k > n$$

and ξ_j, x_j, when passing to homology, will belong to the homology classes Ξ_j and X_j respectively. Set

$$\rho_K \xi_j = \eta_j\,, \qquad \rho_X x_j = y_j\,,$$

Then, since $f^A \eta_j \sim y_j$ for each j by construction of f, we can assume from degree considerations that

$$f^A \eta_j = y_j\,.$$

Since $P_{n,r}$ is contractible, we necessarily have

(7) $\quad g'^A \eta_j = d\alpha_j\,,$

for some $\alpha_j \in A^*(P)$. Let

(8) $\quad \tilde{j}_0^A \alpha_j = \mu_j \in A^*(K_{n-1,r})\,.$

Then

$$d\mu_j = d\tilde{j}_0^A \alpha_j = \tilde{j}_0^A d\alpha_j$$
$$= \tilde{j}_0^A g'^A \eta_j = g_0^A j_0^A \eta_j$$

so that

(9) $\quad c\mu_j = 0\,.$

The equations (7)-(9) show that μ_j is a cycle whose class is transgressive in the fibration ($\mathcal{K}$) to the class of η_j, or the class of ξ_j, when passing to homology. Moreover, it is well-known that $H^*(K_{n-1,r})$, and hence also $I^*(K_{n-1,r})$, are freely generated by the classes of μ_j, so that the fibration ($\mathcal{K}$) is totally transgressive in the sense of § VI.5 , which will be explained in more details, later on.

Now, set

$$\tilde{f}^A \alpha_j = a_j \in A^*(\tilde{X}).$$

Then, we see that

$$\tilde{j}^A a_j = u_j \, (= \mu_j) \in A^*(K_{n-1,r}) ,$$
$$da_j = g^A y_j ,$$
$$du_j = 0 .$$

This shows that u_j is a cycle whose class is transgressive in the fibration $(\mathcal{F})$ to the class of y_j, or the class of x_j, when passing to homology, so that $(\mathcal{F})$ is also totally transgressive, with each generator u_j of

$$I^*(K_{n-1,r}) \approx \text{Free}\,(u_j)$$

transgressive, with the cycle x_j of $I^*(X)$ as its transgression. By theorem in § VI.5 we therefore have

$$I^*(\tilde{X}) \approx \min\,(I^*(X) \underset{\tau}{\otimes} I^*(K_{n-1,r})) .$$

In the formula $\underset{\tau}{\otimes}$ means that the tensor product is to take on some twisted differential

$$d = d_{\otimes} + d_{\tau}$$

where $d_{\otimes}$ is the usual differential as it is in a tensor product, while d_{τ} is the twisted part given by

$$d_{\tau} u_j = x_j .$$

This shows that $I^*(X) \underset{\tau}{\otimes} I^*(K_{n-1,r})$ is just the Cartan-Serre extension of $I^*(X)$ and thus, the theorem is proved.

IV.3 THE CARTAN-SERRE TOWER OF A SPACE AND THE HUREWICZ HOMOMORPHISM

Definition: For any space X of finite type let X_2 be the universal covering space of X, X_3 be the Cartan-Serre extension of X_2 and in general for each

$n \geq 2$, X_{n+1} will be the Cartan-Serre extension of X_n if X_n is a space of type n. Otherwise we simply put $X_{n+1} = X_n$ if X_n is of type $\geq n+1$. Then, we get a sequence of spaces and projections

$$\to X_{n+1} \to X_n \to \dots \to X_2 \to X$$

which will be called the <u>Cartan-Serre tower</u> of the space X.

It is clear that the space X_2 is of finite type and of type ≥ 2 with

$$\pi_s(X_2) \otimes \underline{Q} \approx \pi_s(X) \otimes \underline{Q} \quad \text{for } s \geq 2 .$$

If X is simply connected so that $X_2 = X$ then by Prop. of § IV.2, we see that

$$q_s(X) = q_s(I^*(X)) \quad \text{for } s \geq 2 .$$

In other words, we have the following

<u>Theorem 1</u>: The Hurewicz numbers of a simply connected space of finite type are equal to the Hurewicz numbers of its I^*-measure.

The above theorem can be made more precise to the following

<u>Theorem 2</u>: For a simply connected space X of finite type we have for $n \geq 2$:

$$\mathrm{Vect}_n\, I^*(X) \approx \mathrm{Hom}_{\underline{Q}}(\pi_n(X), \underline{Q}) .$$

<u>Proof</u>: The space X, being simply connected, has its I^*-measure of the form

$$I^*(X) \approx I^*(X_2) \approx \mathrm{Free}\,(x^{(2)}_{j(2)}, \dots, x^{(2)}_{j(n)}, \dots) ,$$

in which

$$\deg x^{(2)}_{j(n)} = n ,$$

and $j(n)$ runs over some index set $J(n)$. It is sufficient to consider the case $J(n) \neq \emptyset$.

Denote by p_n the projection

$$X_n \to X_2 = X$$

in the Cartan-Serre tower of X. Then, by Prop. of §IV.2, we should have

$$I^*(X_n) \approx \mathrm{Free}\,(x^{(n)}_{j(n)},\ x^{(n)}_{j(n+1)},\ldots)$$

with $\deg x^{(n)}_{j(N)} = N$, and $j(N)$ running over the same index sets $J(N)$ as before for $N \geq n$. Moreover, we can take an associated DGA-morphism

$$p^I_n : I^*(X) \to I^*(X_n)$$

induced by p_n such that

$$p^I_n\,(x^{(2)}_{j(N)}) = x^{(n)}_{j(N)} \qquad \text{for } N \geq n\,,$$

while

$$p^I_n(x^{(2)}_{j(m)}) = 0 \qquad \text{for } m < n\,.$$

Let $V^{(n)}_m$ be the vector space spanned by $(x^{(n)}_{j(m)})$ of $I^*(X_n)$, i.e.,

$$V^{(n)}_m = \mathrm{Vect}_m(I^*(X_n))\,,$$

then we have

$$p^I_n : V^{(2)}_n \approx V^{(n)}_n\,.$$

Now we shall define a morphism of vector spaces

$$\tau : V^{(n)}_n \to \mathrm{Hom}_{\underline{Q}}\,(\pi_n(X), Q)$$

as follows.

Consider any element $\alpha \in \pi_n(X) \otimes \underline{Q}$. Since $p_{n\pi} : \pi_n(X_n) \otimes \underline{Q} \approx \pi_n(X) \otimes Q$, there will be an element $\alpha_n \in \pi_n(X) \otimes \underline{Q}$ and maps $f_n : S^n \to X_n$ and $f : S^n \to X$ belonging to α_n and α such that

$$p_{n\pi}(\alpha_n) = \alpha$$

with the following diagram of maps being commutative

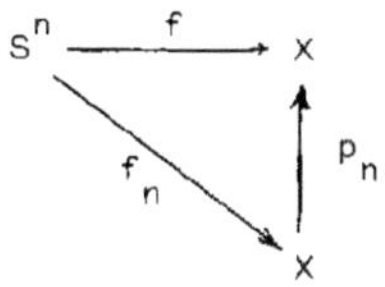

Consider now the following diagram of morphisms:

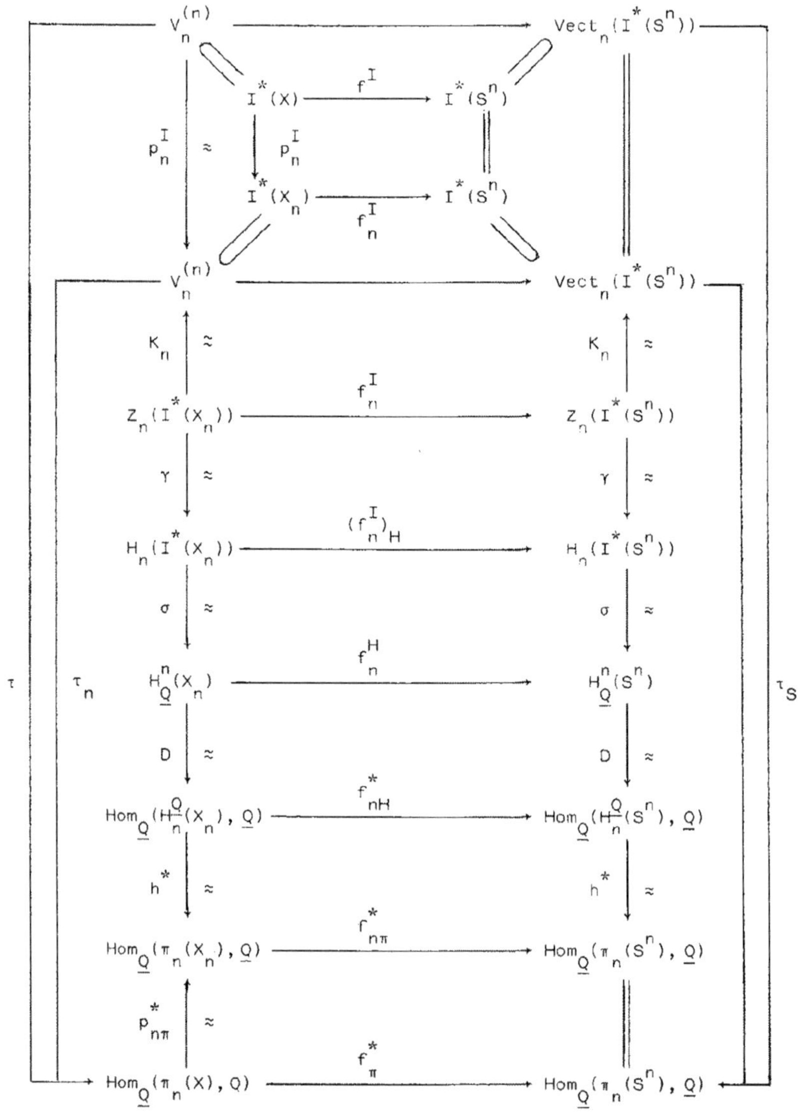

In the diagram, Z_n means groups of cycles of degree n, γ the natural morphism in passing from cycles to their classes, D the duality-morphism, h the Hurewicz morphism, K_n the inclusion, σ the deRham-Sullivan isomorphism, and $*$ the induced dual morphism for Hom. Now, these morphisms are in fact all isomorphisms, so, defining τ_n and τ by

$$\tau_n = (p^*_{n\pi})^{-1} h^* D\sigma\gamma K_n^{-1} ,$$
$$\tau = \tau_n \rho_n^I ,$$

then, τ will be an isomorphism

$$\tau : \mathrm{Vect}_n (I^*(X)) \approx \mathrm{Hom}_{\underline{Q}} (\pi_n(X), \underline{Q}) ,$$

as to be proved. Moreover, since the diagram is clearly commutative, the isomorphism can be made more explicit, viz.,

$$\tau\, x^{(2)}_{j(n)}(\alpha) = \tau_s f^I x^{(2)}_{j(n)}(\iota_n) = (f^I x^{(n)}_{j(n)}, \xi_n) ,$$

in which ι_n is the generator of $\pi_n(S^n)$ corresponding to the identity map of S^n, and ξ_n the generator of degree n in $I^*(S^n)$ arising from $h\iota_n \in H_n^{\underline{Q}}(S^n)$.

The following theorem makes also explicit the Hurewicz homomorphism

$$h : \pi_n(X) \to H_n^{\underline{Q}}(X) .$$

<u>Theorem 3</u>: For a simply connected space X of finite type let Z_n be the group of non-decomposable cycles of degree n of $I^*(X)$, i.e., cycles belonging to $\mathrm{Vect}_n(I^*(X))$, then we have for the Hurewicz homomorphism h:

$$\mathrm{Im}\, h \approx Z_n .$$

<u>Proof</u>: Let us take a basis

$$\mathrm{Vect}_n(I^*(X)) = \mathrm{Vect}\,(z_j^{(n)}, v_k^{(n)})$$

such that each $z_j^{(n)}$ is a cycle while no linear combination of $v_k^{(n)}$ is a cycle. Then we have a basis

$$H^n_Q(X) \approx H_n(I^*(X)) \approx \mathrm{Vect}\,(\mathrm{cl}\, z_j^{(n)}, \mathrm{cl}\, w_1^{(n)}) ,$$

in which $w_1^{(n)}$ are decomposable cycles of degree n in $I^*(X)$, cl denotes the class in passing to homology and Vect the corresponding vector space.

For the isomorphism

$$\tau : \mathrm{Vect}_n(I^*(X)) \approx \mathrm{Hom}_{\underline{Q}}(\pi_n(X), Q)$$

let

$$\tau(z_j^{(n)}) = \alpha_j^* , \qquad \tau(v_k^{(n)}) = \beta_k^* .$$

Then, (α_j^*, β_k^*) will be a basis of $\mathrm{Hom}_{\underline{Q}}(\pi_n(X), \underline{Q})$. With respect to this basis let the dual basis of $\pi_n(X) \otimes \underline{Q}$ be denoted by (α_j, β_k). Take any map $f : S^n \to X$ belonging to β_k. Then, from the definition of τ we get

$$\mathrm{cl}\, w_1^{(n)}(h\beta_k) = \tau_S f^I w_1^{(n)}(\iota_n) = 0 ,$$

since $f^I w_1^{(n)} = 0$, $w_1^{(n)}$ being decomposable. We also have $\mathrm{cl}\, z_j^{(n)}(h\beta_k) = \alpha_j^*(\beta_k) = 0$. Hence, $h\beta_k = 0$. On the other hand $\mathrm{cl}\, z_j^{(n)}(h\alpha_{j'}) = \delta_{j'}$, so that

$$h(\pi_n(X) \otimes \underline{Q}) \approx \mathrm{Vect}(h\alpha_j) \approx \mathrm{Vect}(z_j^{(n)}) \approx Z_n$$

as to be proved.

IV.4 WHITEHEAD PRODUCTS OF HOMOTOPY GROUPS

First, we shall recall the definition of Whitehead products of homotopy groups of a space X to be denoted by

$$[\pi_m(X), \pi_n(X)] \to \pi_{m+n+1}(X) .$$

For this purpose let S^m, S^n be spheres of dimensions m,n and $S^m \vee S^n$ be their union with certain reference points $O_m \in S^m$, $O_n \in S^n$ identified to a single point O_s. Denote the inclusion maps of S^m, S^n into $S^m \vee S^n$ by i_m, i_n respectively. Let ι_m, ι_n be canonical generators of $\pi_m(S^m)$, $\pi_n(S^n)$ as subgroups of $\pi_*(S^m \vee S^n)$. Then, their image in $\pi_*(S^m \vee S^n)$ will be denoted respectively by

$$\tilde{\iota}_m = i_{m\pi}\iota_m , \qquad \tilde{\iota}_n = i_{n\pi}\iota_n .$$

Consider now cubes D^m, D^n of dimensions m,n respectively, for which the boundary of their topological product is given by

$$(D^m \times D^n)^{\bullet} = (D^m \times \dot{D}^n) \cup (\dot{D}^m \times D^n) .$$

Denote the projections of $D^m \times D^n$ onto D^m, D^n by p_m, p_n, the inclusion of

$(D^m \times D^n)^\bullet$ into $D^m \times D^n$ by j, and the shrinkings $(D^m, \dot{D}^m)$ to (S^m, O_m) and $(D^n, \dot{D}^n)$ to (S^n, O_n) by ∂_m, ∂_n. Then we have a diagram of maps below:

$$(D_1) \qquad (D^m \times D^n)^\bullet \xrightarrow{\ j\ } D^m \times D^n \begin{array}{l} \nearrow^{p_m}\ D^m \xrightarrow{\partial_m} S^m\ \searrow^{i_m} \\ \searrow_{p_n}\ D^n \xrightarrow{\partial_n} S^n\ \nearrow_{i_n} \end{array} S^m \vee S^n$$

Define now a map

$$h_0 : (D^m \times D^n)^\bullet \to S^m \vee S^n$$

by

$$h_0 = \begin{cases} i_m \partial_m p_m j \mid D^m \times \dot{D}^n , \\ i_n \partial_n p_n j \mid \dot{D}^m \times D^n . \end{cases}$$

The element $\mathrm{cl}\, h_0$ of $\pi_{m+n-1}(S^m \vee S^n)$ determined by the map h_0 is then the so-called <u>Whitehead product</u> of the elements ι_m and ι_n denoted, as usual, by

$$\mathrm{cl}\, h_0 = [\iota_m, \iota_n] .$$

More generally, for any arcwise-connected space X with certain reference point O consider any elements

$$u \in \pi_m(X) , \qquad v \in \pi_n(X) .$$

Take arbitrary maps

$$f : (S^m, O_m) \to (X,O) ,$$

$$g : (S^n, O_n) \to (X,O)$$

belonging to u, v respectively, and let

$$f \vee g : (S^m \vee S^n, O_s) \to (X,O)$$

be the map defined by

$$f \vee g = \begin{cases} f \mid S^m , \\ g \mid S^n . \end{cases}$$

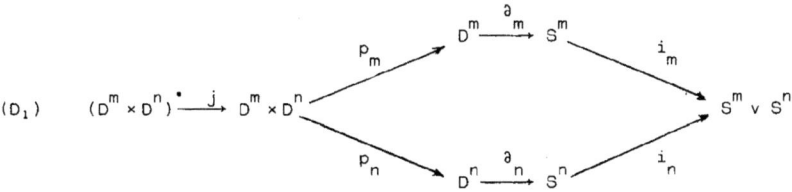

The map

$$(f \vee g)h_0 : (D^m \times D^n)^{\bullet} \to S^m \vee S^n \to X$$

defines then uniquely an element of $\pi_{m+n-1}(X)$ independent of the maps f, g chosen in the elements $u, v \in \pi_n(X)$. So, we can lay down the following

<u>Definition</u>: The element of $\pi_{m+n-1}(X)$ determined by any map $(f \vee g)h_0$ as above is called the <u>Whitehead product</u> of the elements u and v to be denoted by

$$[u,v] \in \pi_{m+n-1}(X) .$$

Now, suppose that X is a simply-connected space of finite type and let

$$\tau : I^*(X) \approx \mathrm{Hom}_{\underline{Q}}(\pi_*(X), \underline{Q})$$

be the isomorphism as graded-vector space as described in §IV.3 . For $m,n \geq 2$ let

$$V_m = \mathrm{Vect}_m(I^*(X)) = \mathrm{Vect}(x_i) ,$$

$$V_n = \mathrm{Vect}_n(I^*(X)) = \mathrm{Vect}(y_j) ,$$

$$V_{m+n-1} = \mathrm{Vect}_{m+n-1}(I^*(X)) = \mathrm{Vect}(z_k)$$

with i, j, k running over certain index sets. For the case m = n it is to be understood that $V_m \equiv V_n$ with (x_i) coincident with (y_j).

Let

$$dz_k = \Sigma a_k^{ij} x_i y_j + \text{other terms},$$

so that $\Sigma a_k^{ij} x_i y_j$ is the quadratic part of dz_k involving elements of degrees m, n. Then, the main object of this section is to prove the following theorem, due to Sulivan [1]:

<u>Theorem</u>: The Whitehead product of the simply-connected space of finite type is completely determined by the quadratic part of the differential of the I^*-measure $I^*(X)$ of X. More precisely, for any elements

$$u \in \pi_m(X) , \qquad v \in \pi_n(X) ,$$

their Whitehead product $[u,v]$ is determined by

$$\tau z_k([u,v]) = - \Sigma a_k^{ij} \tau x_i(u) \cdot \tau y_j(v) .$$

For the proof we shall first make some preparations. Consider thus the diagram (D_1) again and let us realize the spaces and maps as simplicial ones, so that we have the following diagram (D_2) of de-Rham-Sullivan measures A^* and induced DGA-morphisms:

(D_2)

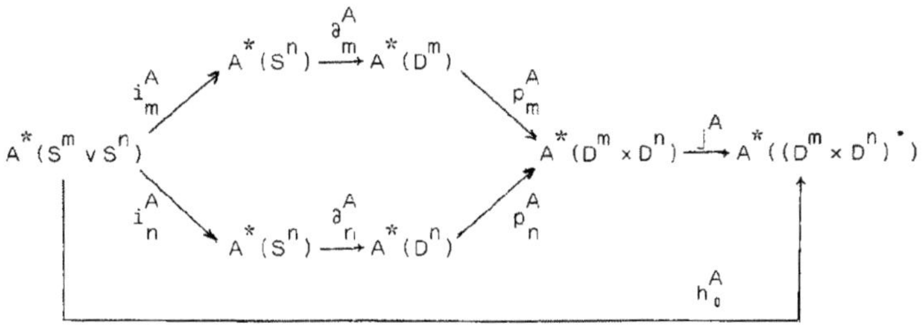

Define now

$$\xi, \eta \in A^*(S^m \vee S^n)$$

by

$$d\xi = 0 \,, \qquad d\eta = 0 \,,$$

$$\deg \xi = m \,, \qquad \deg \eta = n \,,$$

$$\xi(i_m S^m) = 1 \,, \qquad \xi \mid i_n S^n = 0 \,,$$

$$\eta \mid i_m S^m = 0 \,, \qquad \eta(i_n S^n) = 1 \,.$$

As $A^*(D^m)$ and $A^*(D^n)$ are homologically trivial, so we have

$$\partial_m^A i_m^A \xi = d\alpha \,,$$

$$\partial_n^A i_n^A \eta = d\beta$$

for some elements $\alpha \in A^*(D^m)$, $\beta \in A^*(D^n)$, with $\deg \alpha = m-1$, $\deg \beta = n-1$.

Lemma 1: For the element

$$\gamma = j^A(p_m^A \alpha \bullet p_n^A d\beta) \in A^{m+n-1}((D^m \times D^n)^\bullet)$$

we have

$$\gamma((D^m \times D^n)^\bullet) = 1 \,.$$

Proof: By direct calculation we have

$$\gamma((D^m \times D^n)^\bullet) = \int_{(D^m\times D^n)^\bullet} \gamma$$

$$= \int_{j(D^m\times D^n)^\bullet} \hat{p}_m^A\alpha \cdot \hat{p}_n^A d\beta = \int_{D^m\times D^n} d(\hat{p}_m^A\alpha \cdot \hat{p}_n^A d\beta)$$

$$= \int_{D^m\times D^n} \hat{p}_m^A \hat{\partial}_m^A \hat{i}_m^A \xi \cdot \hat{p}_n^A \hat{\partial}_n^A \hat{i}_n^A \eta$$

$$= \int_{D^m} \hat{\partial}_m^A \hat{i}_m^A \xi \cdot \int_{D^n} \hat{\partial}_n^A \hat{i}_n^A \eta$$

$$= \int_{\partial_m D^m} \hat{i}_m^A \xi \cdot \int_{\partial_n D^n} \hat{i}_n^A \eta$$

$$= \xi(i_m S^{m-1}) \cdot \eta(i_n S^{n-1})$$

$$= 1 .$$

Let

$$\varphi : I^*((D^m \times D^n)^\bullet) \to A^*((D^m \times D^n)^\bullet) ,$$

$$\psi : I^*(S^m \vee S^n) \to A^*(S^m \vee S^n)$$

be some associated minimal morphisms which both are H-isomorphisms. Then, there will be elements

$$x,y \in I^*(S^m \vee S^n)$$

with

$$\deg x = m , \quad \deg y = n ,$$

such that $I^*(S^m \vee S^n)$ is freely generated by x,y in degrees m,n respectively for $m \neq n$ or freely generated by x,y together in degree m for $m = n$, and moreover

$$\psi(x) \sim \xi , \quad \psi(y) \sim \eta .$$

Replacing ξ,η by some homologous elements we may assume without loss of generality that

$$\varphi(x) = \xi , \quad \varphi(y) = \eta .$$

For the isomorphism

$$\tau_S : I^*(S^m \vee S^n) \simeq \operatorname{Hom}_{\underline{Q}}(\pi_*(S^m \vee S^n), Q)$$

we have then

$$\tau_s x(\tilde{\iota}_m) = 1 , \qquad \tau_s y(\tilde{\iota}_n) = 1 .$$

Furthermore $\varphi(xy) = \varphi(x)\varphi(y) = \xi\eta = 0$ in $A^*(S^m \vee S^n)$ implies that xy in $I^*(S^m \vee S^n)$ so that

$$xy = dz$$

for some $z \in I^*(S^m \vee S^n)$ with $\deg z = m+n-1$.

Take now

$$f \in u \in \pi_m(X) , \qquad g \in v \in \pi_n(X)$$

as before. Then, for induced morphisms

$$f_\pi : \pi_m(S^m) \to \pi_m(X) ,$$

$$g_\pi : \pi_n(S^n) \to \pi_n(X)$$

we have

$$f_\pi \iota_m = u , \qquad g_\pi \iota_n = v .$$

Moreover we have

<u>Lemma 2</u>:

$$(f \vee g)^I x_i = \tau x_i(u) \cdot x ,$$

$$(f \vee g)^I y_j = \tau y_j(v) \cdot y .$$

<u>Proof</u>: From the diagram

$$\begin{array}{ccc} I^*(X) & \xrightarrow[\approx]{\tau} & \mathrm{Hom}_{\underline{Q}}(\pi_m(X), \underline{Q}) \\ {\scriptstyle f^I}\downarrow & & \downarrow{\scriptstyle f_\pi^*} \\ I^*(S^m) & \xrightarrow[\approx]{\tau_s} & \mathrm{Hom}_{\underline{Q}}(\pi_m(S^m), \underline{Q}) \end{array}$$

we get

$$\tau_s f^I x_i(\iota_m) = \tau x_i(f_\pi \iota_m) = \tau x_i(u) .$$

Similarly

$$\tau_s g^I y_j(\iota_m) = \tau y_j(v) .$$

Consider now the diagram

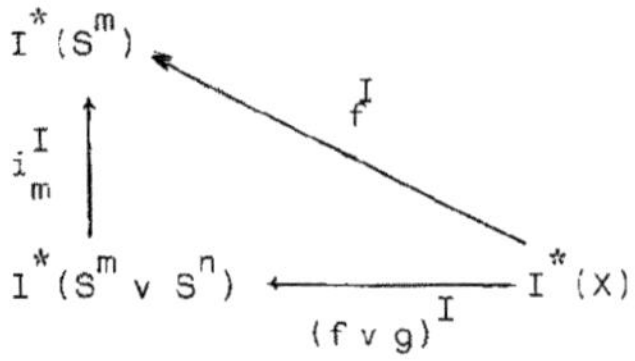

We have then

$$\tau_s (f \vee g)^I x_i(\tilde{\iota}_m)$$
$$= \tau_s f^I x_i(\iota_m) = \tau x_i(u)$$
$$= \tau x_i(u) \cdot \tau_s x(\tilde{\iota}_m) \ .$$

Whence

$$(f \vee g)^I x_i = \tau x_i(u) \cdot x \ .$$

Similarly, for $(f \vee g)^I y_j$, as to be proved.

Now, according to §II.4, there will be, in writing

$$\mathcal{A} = A^*((D^m \times D^n)^\bullet) , \qquad \mathcal{B} = I^*((D^m \times D^n)^\bullet)$$

for simplicity, a commutative diagram of DGA-morphisms arising from h_0 as below:

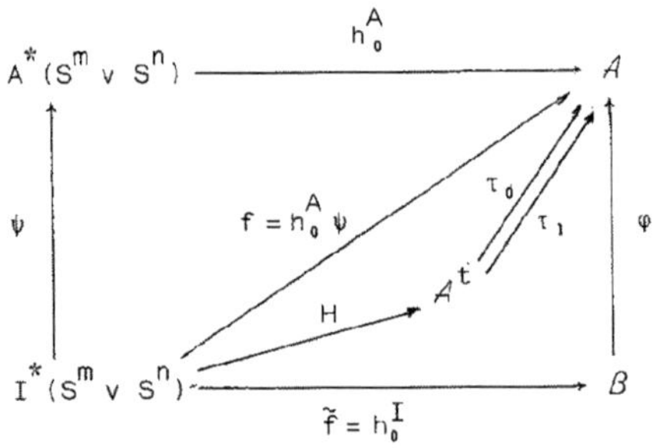

Let us take $w \in \mathcal{B}$ such that $\deg w = m+n-1$, $dw = 0$, and $\varphi w \sim \gamma$. Then it i clear by Lemma 1 that, for ι_{m+n-1} as the canonical generator of $\pi_{m+n-1}((D^m \times D^n)^\bullet)$,

$$\tau_s w(\iota_{m+n-1}) = 1 \ .$$

Lemma 3: The induced DGA-morphism h_0^I is such that

$$h_0^I z = -w .$$

Proof: With h_0^A, φ, ψ and hence $f = h_0^A \psi$ supposed given, we can determine stepwise h_0^I and H according to the procedure described in the proof of the Lifting Lemma of §II.4, in the following manner.

Set

$$f = h_0^A \psi \quad , \qquad \tilde{f} = h_0^I .$$

Then, we have

$$fx = dj^A p_m^A \alpha \sim 0 \quad \text{in} \quad \mathcal{A} ,$$

$$fy = dj^A p_n^A \beta \sim 0 \quad \text{in} \quad \mathcal{A} ,$$

and also

$$\tilde{f}x = 0 , \qquad \tilde{f}y = 0 .$$

According to the procedure described in §II.4 we have then

$$Hx = fx - d(tj^A p_m^A \alpha) ,$$

$$Hy = fy - d(tj^A p_n^A \beta) .$$

Proceeding further, we have

$$\tilde{f}dz = \tilde{f}(xy) = \tilde{f}x \cdot \tilde{f}y = 0 = db ,$$

where b may be simply taken to be 0. Then

$$\begin{aligned}
\varphi b - fz - IHdz &= -IH(xy) = -I(Hx \cdot Hy) \\
&= -I[(fx - d(tj^A p_m^A \alpha))(fy - d(tj^A p_n^A \beta))] \\
&= -I[(fx - tdj^A p_m^A \alpha - dt \cdot j^A p_m^A \alpha)(fy - tdj^A p_n^A \beta - dt \cdot j^A p_n^A \beta)] \\
&= -I[(1-t)^2 dj^A p_m^A \alpha \cdot dj^A p_n^A \beta - (1-t) \cdot dt \cdot j^A p_m^A \alpha \cdot dj^A p_n^A \beta - \\
&\qquad - (-1)^{\deg \alpha + 1} \cdot (1-t) dt \cdot dj^A p_m^A \alpha \cdot j^A p_n^A \beta] \\
&= \int_0^1 (1-t)(j^A p_m^A \alpha \cdot dj^A p_n^A \beta + (-1)^{\deg \alpha + 1} \cdot dj^A p_m^A \alpha \cdot j^A p_n^A \beta) dt \\
&= \frac{1}{2}(j^A p_m^A \alpha \cdot dj^A p_n^A \beta + (-1)^{\deg \alpha + 1} \cdot dj^A p_m^A \alpha \cdot j^A p_n^A \beta) =
\end{aligned}$$

$$= j^A(p_m^A \alpha \cdot dp_n^A \beta) - \frac{1}{2}(-1)^{\deg \alpha} \cdot dj^A(p_m^A \alpha \cdot p_n^A \beta)$$

$$\sim \varphi w \quad \text{in} \quad \mathcal{A} .$$

According to §II.4, we may then take

$$\tilde{f} z = b - w = -w ,$$

i.e.,

$$h_0^I z = -w ,$$

as to be proved.

We are now in a position to prove the Theorem about the Whitehead product, as follows.

Proof of Theorem: Use the same notations as before with

$$f \in u \in \pi_m(X)$$

and

$$g \in v \in \pi_n(X) .$$

We have then the following diagram of maps with $h = (f \vee g)h_0$:

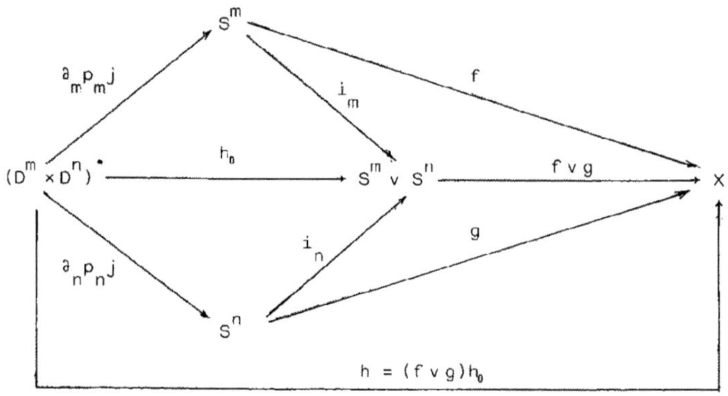

It follows that

$$d(f \vee g)^I z_k = (f \vee g)^I dz_k = \Sigma\, a_k^{ij} (f \vee g)^I x_i y_j ,$$

the terms in dz_k besides the quadratic part in x, y make no contributions owing to degree considerations. By Lemma 2 we have therefore

$$
\begin{aligned}
d(f \vee g)^{I} z_k &= \Sigma a_k^{ij}\, \tau x_i(u) \cdot \tau y_j(v) \cdot xy \\
&= \Sigma a_k^{ij}\, \tau x_i(u) \cdot \tau y_j(v) \cdot dz ,
\end{aligned}
$$

so that

$$(f \vee g)^{I} z_k - \Sigma a_k^{ij}\, \tau x_i(u) \cdot \tau y_j(v) \cdot z$$

is a cycle in $I^*(S^m \vee S^n)$. Since $I^*(S^m \vee S^n)$ has no non-zero cycles in degree $m+n-1$ we get then

$$(f \vee g)^{I} z_k = \Sigma a_k^{ij}\, \tau x_i(u) \cdot \tau y_j(v) \cdot z .$$

Whence

$$
\begin{aligned}
h^{I} z_k &= h_0^{I}(f \vee g)^{I} z_k \\
&= \Sigma a_k^{ij}\, \tau x_i(u) \cdot \tau y_j(v) \cdot h_0^{I} z .
\end{aligned}
$$

Now, by definition we have

$$h_{0\pi}\, \iota_{m+n-1} = [\tilde{\iota}_m, \tilde{\iota}_n] ,$$

and

$$h_{\pi}\, \iota_{m+n-1} = (f \vee g)_{\pi}[\tilde{\iota}_m, \tilde{\iota}_n] = [u,v] .$$

Consequently

$$
\begin{aligned}
\tau z_k([u,v]) &= \tau z_k(h_{\pi}\, \iota_{m+n-1}) \\
&= \tau_s\, h^{I} z_k(\iota_{m+n-1}) \\
&= \Sigma\, a_k^{ij}\, \tau x_i(u) \cdot \tau y_j(v) \cdot \tau_s\, h_0^{I} z(\iota_{m+n-1}) \\
&= -\Sigma a_k^{ij}\, \tau x_i(u) \cdot \tau y_j(v) .
\end{aligned}
$$

This completes the proof of the theorem.

Chapter V

I^* - MEASURE OF A HOMOGENEOUS SPACE - THE CARTAN THEOREM

V.1 DGA OF LEFT-INVARIANT FORMS ON A LIE GROUP

We shall recapitulate some well-known facts about the DGA of left-invariant forms on a Lie group to be used in the present Chapter. Those proofs which can be found in standard textbooks on Lie groups or related subjects will be omitted. To fix the ideas, we shall adopt the following notations held throughout the whole chapter.

G is a connected Lie group of dimension r.

$\underline{g}$ is the Lie algebra of G, formed of all left-invariant vector fields on G.

$i, j, k, \ldots, =$ indices with range in $1, 2, \ldots, r$.

L_g (resp. R_g) for any $g \in G$ is the left- (resp. right-) translation of G defined by

$$L_g x = gx, \qquad (x \in G)$$

$$(\text{resp. } R_g x = xg)$$

Int_g for any $g \in G$ is the <u>interior transformation</u> of G defined by

$$\text{Int}_g = L_g R_{g^{-1}} = R_{g^{-1}} L_g ,$$

or

$$\text{Int}_g x = gxg^{-1}, \qquad x \in G .$$

X_i $(i=1, \ldots, r)$ is a fixed basis of $\underline{g}$ with Lie brackets $[\,,]$ and structural constants c_{ij}^k verifying the following conditions:

$$[X_i , X_j] = c_{ij}^k X_k ,$$

$$c_{ij}^k = - c_{ji}^k ,$$

$$c_{is}^p c_{jk}^s + c_{js}^p c_{ki}^s + c_{ks}^p c_{ij}^s = 0 .$$

$A^*(G)$ is the DGA of all differential forms on G.

$A^L(G)$ is the subalgebra of $A^*(G)$ formed of all left-invariant differential forms on G, i.e., forms α for which

$$L_g^* \alpha = \alpha$$

for any $g \in G$.

$\{\omega^i\}$ form the basis of left-invariant forms dual to $\{X_i\}$, $(i=1, \ldots, r)$:

$$\omega^i(X_j) = \delta_{ij} .$$

Then, $A^L(G)$ is a free exterior DGA on generators ω^i with degree

$$\deg \omega^i = 1 .$$

The differential will satisfy the following Maurer-Cartan equations

$$d\omega^k = -\frac{1}{2} c^k_{ij} \omega^i \wedge \omega^j ,$$

in which $\wedge$ means the usual exterior multiplication occasionally omitted in what follows.

Ad is the adjoint representation of G defined by

$$Ad : \begin{cases} G \to GL(\underline{g}) , \\ g \to (Int_g)_* . \end{cases}$$

As usual, we denote for any $g \in G$ the corresponding general linear transformation of $GL(\underline{g})$ by $Ad_{\underline{g}}$ instead of $Ad(g)$, viz.

$$Ad_g = Int_{g*} : \underline{g} \to \underline{g} .$$

ad is the adjoint representation of $\underline{g}$ on $GL(\underline{g})$ induced by Ad ,

$$ad = Ad_* : \underline{g} \to gl(\underline{g}) .$$

Again, as usual, we denote for any $X \in g$, the corresponding linear transformation of $gl(\underline{g})$ by ad_X instead of $ad(X)$.

For any $g \in G$ the interior transformation $Int_{g^{-1}}$ will induce a DGA-automorphism of $A^*(G)$ which will also be denoted as

$$Ad_g = Int^*_{g^{-1}} .$$

Then

$$\mathrm{Ad} : \begin{cases} G \to \text{DGA-Aut}\, A^*(G) \ , \\ g \to \mathrm{Ad}_g \end{cases}$$

is a group-morphism which will induce for any $X \in \underline{g}$ a <u>derivation</u> ad_X of $A^*(G)$. The term <u>derivation</u> means that for any differential form $\alpha, \beta \in A^*(G)$ we have

$$\mathrm{ad}_X(\alpha \wedge \beta) = \mathrm{ad}_X \alpha \wedge \beta + \alpha \wedge \mathrm{ad}_X \beta \ .$$

Moreover, for any $g \in G$, Ad_g is not only a DGA-automorphism of $A^*(G)$, but also one of $A^L(G)$, so that for any $X \in g$, ad_X is not only a derivation of $A^*(G)$, but also one of $A^L(G)$.

With respect to the chosen basis $\{X_i\}$ of g and its dual basis $\{\omega^i\}$ of $A^L(G)$ let us write for simplicity

$$\mathrm{ad}_i = \mathrm{ad}_{X_i} \ .$$

Then, we have the following properties:

P 1. For any $X,Y \in \underline{g}$,

$$\mathrm{ad}_X Y = [X,Y] \ .$$

P 2. $$\mathrm{ad}_i\, \omega^k = -c^k_{ij}\, \omega^i \ .$$

P 3. For any $\alpha \in A^L(G)$,

$$d\alpha = \frac{1}{2}\omega^i \wedge \mathrm{ad}_i\, \alpha \ .$$

P 4. $$d\, \mathrm{ad}_i = \mathrm{ad}_i\, d \quad \text{on} \quad A^L(G) \ .$$

P 5. For any $X,Y \in g$ and any $\omega \in A^L(G)$ with $\deg \omega = 1$,

$$(\mathrm{ad}_X\, \omega)Y = -\omega(\mathrm{ad}_X Y) \ .$$

The Lie algebra $\underline{g}$ of a compact Lie group G is called a <u>compact Lie algebra</u>. It is characterized by possessing a positive definite euclidean metric (,) on $\underline{g}$ which is invariant under Ad_g for any $g \in G$:

$$(\mathrm{Ad}_g X, \mathrm{Ad}_g Y) = (X,Y) , \qquad X,Y \in \underline{g} .$$

This implies that for any $X,Y,Z \in \underline{g}$:

$$(\mathrm{ad}_Z X, Y) + (X, \mathrm{ad}_Z Y) = 0 .$$

In what follows, G will now be specialized to be a compact Lie group with its compact Lie algebra $\underline{g}$. In this case, invariant integration on G may then be defined and will be used freely in what follows. The basis $\{X_i\}$ will also be specialized to be an orthonormal one with respect to the Ad-invariant positive definite metric $(\,,\,)$ in $\underline{g}$, with $\{\omega^i\}$ the dual basis to $\{X_i\}$. For any forms in $A^L(G)$,

$$\alpha = a_{i_1 \ldots i_s} \omega^{i_1} \wedge \ldots \wedge \omega^{i_s} ,$$
$$\beta = b_{j_1 \ldots j_s} \omega^{i_1} \wedge \ldots \wedge \omega^{i_s}$$

of same degree s, we shall set as definition:

$$(\alpha, \beta) = a_{i_1 \ldots i_s} b_{j_1 \ldots j_s} \delta^{i_1 \ldots i_s}_{j_1 \ldots j_s} ,$$

in which $\delta^{i_1 \ldots i_s}_{j_1 \ldots j_s}$ is $+1$ or -1 according to $(j_1, \ldots, j_s)$ being an even or odd permutation of $(i_1, \ldots, i_s)$ and otherwise 0.

A simple consequence of the existence of an adjoint-invariant positive definite metric is the following:

With respect to an orthonormal basis $\{X_i\}$ of $\underline{g}$ the structural constants c^k_{ij} are skew-symmetric in all 3 indices i,j,k.

Besides the usual differential operator d, let us introduce now some other linear operators in the DGA $A^L(G)$ for the compact Lie group G :

O 1. For any

$$\alpha = \omega^i \wedge \beta$$

with β not involving ω^i, we define

$$\partial^i \alpha = \beta ,$$

which is $\frac{\partial \alpha}{\partial \omega^i}$ in the old notation of E. Cartan.

0 2. For any α we define

$$\delta^i \alpha = \omega^i \wedge \alpha .$$

0 3. For any α of degree $s > 0$ we define a form $\partial\alpha$ of degree $s-1$ by

$$(\partial\alpha, \beta) = (\alpha, d\beta)$$

for any β of degree $s-1$.

Below, we list a few of the properties about these operators for which the proofs are all quite simple $(\alpha, \beta \in A^L(G)$ with due degrees$)$:

$$d\alpha = \frac{1}{2} \delta^i \mathrm{ad}_i \alpha ,$$

$$\partial\alpha = -\frac{1}{2} \partial^i \mathrm{ad}_i \alpha ,$$

$$\mathrm{ad}_i \omega^i = 0 \quad \text{(no summation)} ,$$

$$\mathrm{ad}_i \omega^k = -c^k_{ij} \omega^j ,$$

$$\mathrm{ad}_i (\alpha \wedge \beta) = \mathrm{ad}_i \alpha \wedge \beta + \alpha \wedge \mathrm{ad}_i \beta ,$$

$$d(\alpha \wedge \beta) = d\alpha \wedge \beta + (-1)^{\deg \alpha} \cdot \alpha \wedge d\beta ,$$

$$\partial^i (\alpha \wedge \beta) = \partial^i \alpha \wedge \beta + (-1)^{\deg \alpha} \cdot \alpha \wedge \partial^i \beta ,$$

$$\delta^i (\alpha \wedge \beta) = \delta^i \alpha \wedge \beta + (-1)^{\deg \alpha} \cdot \alpha \wedge \delta^i \beta ,$$

$$(\alpha, \partial^k \beta) = (\delta^k \alpha, \beta) ,$$

$$(\partial\alpha, \beta) = (\alpha, d\beta) ,$$

$$(\mathrm{ad}_i \alpha, \beta) + (\alpha, \mathrm{ad}_i \beta) = 0 .$$

Remark: $\partial(\alpha \wedge \beta) \neq \partial\alpha \wedge \beta \pm \alpha \wedge \partial\beta$ in general.

In the vector space $A^{L,s}(G)$, spanned by forms of degree s in $A^L(G)$, we shall denote by $\mathrm{Im}^s d$ the subspace spanned by all these elements in the image of d. Similarly, for $\mathrm{Im}^s \partial$, $\mathrm{Ker}^s \delta$, and $\mathrm{Ker}^s \partial$. Then, we have the following orthogonal decomposition of the vector space $A^L(G)$:

<u>Theorem 1:</u> The vector space $A^{L,s}(G)$ is the direct sum of 3 subspaces mutually orthogonal to each other, viz.

$$A^{L,s}(G) = \mathrm{Im}^s\, d \oplus \mathrm{Im}^s\, \partial \oplus \mathrm{Ker}^s\, d \cap \mathrm{Ker}^s\, \partial .$$

Moreover,

$$\mathrm{Ker}^s\, d = \mathrm{Im}^s\, d \oplus \mathrm{Ker}^s\, d \cap \mathrm{Ker}^s\, \partial ,$$

$$\mathrm{Ker}^s\, \partial = \mathrm{Im}^s\, \partial \oplus \mathrm{Ker}^s\, d \cap \mathrm{Ker}^s\, \partial .$$

The subspaces appearing in the decomposition are again orthogonal to each other.

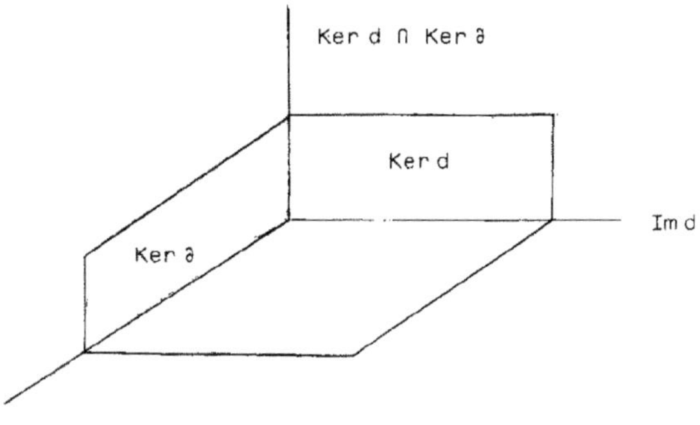

The proof follows directly from the above formulae about the various operators and will be omitted. When no misunderstanding can occur, we shall also simply write $\mathrm{Im}\, d$ instead of $\mathrm{Im}^s\, d$, etc.

<u>Definition:</u> A left-invariant differential form $\alpha \in A^L(G)$ is called simply an <u>invariant form</u> if it is also right-invariant, or what is the same, if any one of the following 3 equivalent conditions is true:

$$R_g^* \alpha = \alpha , \qquad \text{for any } g \in G ,$$

or $$\mathrm{Ad}_g \alpha = \alpha , \qquad \text{for any } g \in G ,$$

or $$\mathrm{ad}_k \alpha = 0 , \qquad \text{for any } k .$$

From the formulae listed before we see that the set of all invariant forms is naturally forming a sub-DGA of $A^L(G)$ and will be denoted henceforth by $A^I(G)$.

<u>Theorem 2:</u> $A^I(G) = \mathrm{Ker}\, d \cap \mathrm{Ker}\, \partial$.

Proof: For any $\gamma \in A^I(G)$ we have by definition $ad_k \gamma = 0$ for any k. Hence, $d\alpha = \frac{1}{2}\delta^k \, ad_k \alpha = 0$ and

$$\partial\alpha = -\frac{1}{2}\partial^k \, ad_k \alpha = 0 \,, \quad \text{or}$$

$$\gamma \in \operatorname{Ker} d \cap \operatorname{Ker} \partial \,.$$

Conversely, let $\gamma \in \operatorname{Ker} d \cap \operatorname{Ker} \partial$.

Then, $\gamma \in \operatorname{Ker} d$ implies

$$\gamma \sim \int_G Ad_g \gamma \, dg = \gamma^I \in A^I(G) \,.$$

By what we have already proved

$$\gamma^I \in \operatorname{Ker} d \cap \operatorname{Ker} \partial \,.$$

Hence, $\gamma \sim \gamma^I$ implies

$$\gamma - \gamma^I \in (\operatorname{Ker} d \cap \operatorname{Ker} \partial) \cap \operatorname{Im} d \,.$$

By Theorem 1 $\operatorname{Im} d$ is orthogonal to $\operatorname{Ker} d \cap \operatorname{Ker} \partial$.
Hence, $\gamma = \gamma^I \in A^I(G)$.

c.e.d.

Notation: The orthogonal projection of $A^L(G)$ onto $A^I(G) = \operatorname{Ker} d \cap \operatorname{Ker} \partial$ wil be denoted by π.

Theorem 3: The projection

$$\pi : \operatorname{Ker} d \to A^I(G)$$

is an algebraic epimorphism.

Proof: For $\alpha, \beta \in \operatorname{Ker} d$ we have by Theorems 1,2

$$\alpha = \pi\alpha + d\alpha' \,, \qquad \beta = \pi\beta + d\beta'$$

for some α', β'. Then,

$$\alpha \wedge \beta = \pi\alpha \wedge \pi\beta + d(\alpha' \wedge \pi\beta \pm \pi\alpha \wedge \beta' + \alpha' \wedge d\beta') \,.$$

Moreover, $\pi\alpha \wedge \pi\beta \in A^I(G)$ since $A^I(G)$ is a DGA.

Consequently

$$\pi\alpha \wedge \pi\beta = \pi(\alpha \wedge \beta)$$

or π is an algebraic morphism. That π is an epimorphism follows directly from Theorem 2.

V.2 HOMOGENEOUS SPACE AND INVARIANT FORMS - METHOD OF E. CARTAN

In what follows, let G be again a connected Lie group of dimension r, as before. Besides, let H be a closed connected subgroup of G and

$$M = G/H$$

be the homogeneous space consisting of all left-cosets of H in G. The natural projection of G to H will be denoted by π, we also suppose $\dim M = n$ so that $\dim H = r - n$.

Now, the following notations will be adopted:

$i, j, k, \ldots$ = indices among $1, 2, \ldots, r$

$a, b, c, \ldots$ = indices among $1, 2, \ldots, n$

$\alpha, \beta, \gamma, \ldots$ = indices among $n+1, \ldots, r$.

For the basis $X_1, \ldots, X_r$ of the Lie algebra g, we shall choose in the way that $X_{n+1}, \ldots, X_r$ at identity e of H and G will span the tangent space of H at e. For the structural constants we have then

$$c^a_{\alpha\beta} = 0 .$$

Now, to each $g \in G$ there is a left-translation T_g of M defined by $T_g(g'H) = gg'H$ for any $g'H \in M$. Thus, G operates transitively on M with H as the isotropy group at 0, the projection of the identity e on M. The following definition is now fundamental:

<u>Definition</u>: A differential form $\alpha \in A^*(M)$ will be called an <u>invariant form</u> if it is invariant under all left-translations T_g, i.e. if $T^*_g\alpha = \alpha$ for all $g \in G$. The collection of all such invariant forms clearly form a sub-DGA of $A^*(M)$ and will be denoted by $A^L(M)$.

The importance of the notion of invariant forms may be seen from the following theorem of E. Cartan:

Theorem 1: If G is compact, then the inclusion of $A^L(M)$ in $A^*(M)$ is an H-isomorphism.

The proof follows simply from the existence of an invariant measure, so that invariant integration is possible for a compact group and will be omitted. We remark only that the theorem is not true in the case of non-compact groups, as seen from the following example given by E. Cartan himself.

Ex. Let G be the group D_N of displacements in an euclidean N-space R^N and H be the subgroup O_N of rotations about a fixed point $0 \in R^N$. Then, the homogeneous space $M = G/H$ is just the euclidean space R^N. With a coordinate system $(x_1, \ldots, x_N)$ about 0 in R^N, it can be easily verified that there exist no invariant forms on R^N in degrees > 0 and $< N$, while there is one $\neq 0$ in the degree N, viz.

$$\omega = dx_1 \wedge \ldots \wedge dx_N .$$

Since $d\omega = 0$ and $\omega \neq d\alpha$ for any invariant form α, the inclusion $A^L(R^N) \subset A^*(R^N)$ cannot be an H-isomorphism.

In a further step, the method of E. Cartan consists in reducing $A^L(M)$ to some DGA's easier to deal with. For this purpose, let us introduce the collection W_B^I of differential forms ξ in $A^*(G)$ verifying the following conditions:

1°. $L_g^* \xi = \xi$ for any $g \in G$,

2°. $R_h^* \xi = \xi$ for any $h \in H$,

where R_h is the right-translation in G defined by $R_h g = gh$ for any $g \in G$.

3°. For any $g \in G$ and any tangent vector $\tilde{X}_1^g, \ldots, \tilde{X}_s^g$ of G at g where $s = \deg \xi$,

$$\xi(X_1^g, \ldots, \tilde{X}_s^g) = 0$$

whenever some of $\tilde{X}_i^g$ lie in the tangent subspace of gH at g.

It is easy to verify that the projection $\pi : G \to M$ will induce a module isomorphism

$$\pi^* : A^L(M) \approx W_B^I .$$

The next step consists then in introducing some differential in W_B^I to turn it into a DGA such that π^* becomes a DGA-isomorphism. In fact, the condition 1° shows that any $\xi \in W_B^I$ of degree s is expressible in the form

$$\xi = x_{i_1 \dots i_s} \omega^{i_1} \wedge \dots \wedge \omega^{i_s}$$

with $x_{i_1 \dots i_s}$ real constants. The condition 3° implies further that any form ω^α cannot occur in the expression of ξ, or ξ should be of the form

$$(1) \qquad \xi = x_{a_1 \dots a_s} \omega^{a_1} \wedge \dots \wedge \delta^{a_s} .$$

Now, the differential d in $A^L(G)$ is determined by the Maurer-Cartan equations

$$d\omega^k = -\frac{1}{2} c^k_{ij} \omega^i \wedge \omega^j ,$$

in particular

$$d\omega^a = -\frac{1}{2} c^a_{bc} \omega^b \wedge \omega^c + \zeta^a ,$$

where

$$\zeta^a = -\frac{1}{2} c^a_{\beta\gamma} \omega^\beta \wedge \omega^\gamma - c^a_{b\gamma} \omega^b \wedge \omega^\gamma .$$

Since

$$\pi^* : A^L(M) \to A^L(G)$$

preserves differential $d\xi \in \pi^* A^L(M) = W_B^I$ too for any $\xi \in \pi^* A^L(M) = W_B^I$. Hence, if we introduce an operator d_B in $A^L(G)$ given by

$$(2) \qquad d_B \omega^a = -\frac{1}{2} c^a_{bc} \omega^b \wedge \omega^c ,$$

then the same $d\xi \in W_B^I$ will be obtained by calculations using d_B instead of d. Note that ω^a itself is in general not in W_B^I. Moreover, let us denote by the same letter d_B the differential in W_B^I induced by the above d_B. Then, the DGA W_B^I with this differential will be in DGA-isomorphism with $A^L(M)$ under π^*. Hence, we have the following

<u>Theorem 2</u>: Let W_B^I be the sub-DGA of $A^L(G)$ consisting of all forms ξ generated by ω^a with coefficients x all real constants, as given by (1), and a differential d_B based on calculations given by (2). Then,

$$\pi^* : A^L(M) \to W_B^I$$

is a DGA-isomorphism.

In combining Theorems 1 and 2, we get the following theorem about the I^*-measure of a homogeneous space arising from a compact Lie group:

Theorem 3: The I^*-measure on a real field of a homogeneous space $M = G/H$ with G connected compact and H closed connected is given by

$$I^*(M) \approx \min W_B^I .$$

Thus, $I^*(M)$ and in particular $H^*_{\underline{R}}(M) = H(I^*(M))$ is completely determined by the structural constants c^a_{bc}.

The above theorem reduces the determination of the I^*- and thus also the $H^*_{\underline{R}}$-measure of a homogeneous space $M = G/H$ with G compact to a purely algebraic problem. A further reduction is possible in the case of the so-called symmetric spaces. We recall first its definition.

Let G be a connected Lie group, as before. An automorphism σ of G is said to be involutive if σ^2= identity (we assume σ itself is not the identity). Let H be the subgroup of G consisting of elements invariant under σ. Then, the homogeneous space $M = G/H$ is called a symmetric space with a symmetry given by $\sigma(gH) = (\sigma g)H$.

Let the indices $i, a, \alpha, \ldots$, etc. be as before. Then, for a symmetric space $M = G/H$ with symmetry σ we can choose the basis $\{X_i\}$ such that the linear transformation $\sigma_* : \underline{g} \to \underline{g}$ induced by σ will be given by

$$\sigma_*(X_a) = -X_a ,$$

$$\sigma_*(X_\alpha) = +X_\alpha .$$

Lemma: With respect to the above basis the structural constants will satisfy the relations:

$$c^c_{ab} = 0 ,$$

$$c^a_{\alpha\beta} = 0 ,$$

and

$$c^\beta_{\alpha a} = 0 .$$

Proof: Let us prove e.g. the first relation. We have in fact

$$\sigma_*[X_a,X_b] = c^k_{ab}\sigma_* X_k = c^\alpha_{ab}X_\alpha - c^c_{ab}X_c ,$$

$$[\sigma_* X_a, \sigma_* X_b] = [-X_{a_1}, -X_b] = c^k_{ab}X_k$$

$$= c^\alpha_{ab}X_\alpha + c^c_{ab}X_c .$$

Since σ_* preserves bracket operations, or $\sigma_*[X_a, X_b] = [\sigma_* X_a, \sigma_* X_b]$, we get by comparison $c^c_{ab} = 0$. The proofs of other relations are similar.

Theorem 4: For a symmetric space $M = G/H$ all invariant forms are closed, i.e., $d\alpha = 0$ for any $\alpha \in A^L(M)$.

Proof: Let us choose the basis $\{X_i\}$ in $\underline{g}$ as above. By Theorem 2, it is sufficient to prove that $d_B\xi = 0$ for any form $\xi \in W^I_B$. Consider therefore any such form

$$\xi = x_{a_1 \dots a_s} \omega^{a_1} \wedge \dots \wedge \omega^{a_s} ,$$

with x all real constants. Now, by the above lemma the equation for which the differential d_B in W^I_B is based becomes

$$d_B \omega^a = -\frac{1}{2} c^a_{bc} \omega^b \wedge \omega^c = 0 .$$

This proves that $d_B = 0$ in W^I_B and the theorem is thus proved.

Combining the above theorem with Theorems 2 and 3, we get then the following

Theorem 5: For a symmetric space $M = G/H$ with G connected compact not only the $H^*_{\underline{R}}$-measure of M is determined by the I^*-measure of M but also conversely the I^*-measure is determined by the $H^*_{\underline{R}}$-measure, viz.

$$I^*(M) \simeq \min H^*_{\underline{R}}(M)$$

($\underline{k} = \underline{R}$ here).

V.3 THE WEIL ALGEBRA

Let G be a compact connected Lie group and H be a closed connected subgroup with notations as before. In the preceding section we have reduced the determination of the I^*-measure (on real field) of the corresponding homogeneous space $M = G/H$ to a purely algebraic problem via some sub-DGA W^I_B of $A^L(G)$ consisting of constant-coefficient forms (henceforth, the symbol $\wedge$ will be dropped).

(1) $$\xi = x_{a_1 \dots a_s} \omega^{a_1} \wedge \dots \wedge \omega^{a_s} = x_{a_1 \dots a_s} \omega^{a_1} \dots \omega^{a_s} ,$$

which are Ad_h-invariant, i.e.,

(2) $$Ad_h \xi = \xi , \qquad \text{for any } h \in H , \quad \text{or}$$

(3) $$ad_\alpha \xi = 0 .$$

The differential d_B in W_B^I is based on the formula

(4) $$d_B \omega^a = -\frac{1}{2} c^a_{bc} \omega^b \omega^c .$$

In fact, H being a subgroup implies $c^a_{\alpha\beta} = 0$. For the differential d in $A^L(G)$ we have therefore

$$d\omega^a = -\frac{1}{2} c^a_{ij} \omega^i \omega^j = -\frac{1}{2} c^a_{bc} \omega^b \omega^c - c^a_{\alpha b} \omega^\alpha \omega^b = d_B \omega^a + \omega^\alpha ad_\alpha \omega^a .$$

For any ξ given as above we have then

$$d\xi = \Sigma (-1)^{k-1} x_{a_1 \dots a_s} \omega^{a_1} \dots (d\omega^{a_k}) \dots \omega^{a_s}$$

$$= \Sigma (-1)^{k-1} x_{a_1 \dots a_s} \omega^{a_1} \dots (d_B \omega^{a_k} + \omega^\alpha ad_\alpha \omega^{a_k}) \dots \omega^{a_s} ,$$

or

$$d\xi = d_B \xi + \omega^\alpha ad_\alpha \xi .$$

For $\xi \in W_B^I$ we have $ad_\alpha \xi = 0$, so that $d_B \xi = d\xi$ is just the differential induced in W_B^I from d in $A^L(G)$.

In this section, we shall reduce the study of I^*-measure of $M = G/H$ with G compact, by one step, from $W_{\bar{B}}^{\bar{I}}$ to some other DGA, viz. the so-called <u>Weil algebra</u> associated to the pair (G,H). First, we introduce the Weil algebra of the Lie group G itself:

<u>Definition</u>: For the Lie group G we define a DGA

(5) $$\text{Weil}\,(G) = \text{Free}\,(\omega^i, \bar{\omega}^i)$$

with

(6) $$\begin{cases} \deg \omega^i = 1 , \\ \deg \bar{\omega}^i = 2 , \end{cases}$$

and differential d_W given by

$$(7)\quad \begin{cases} d_W\,\omega^i = \bar\omega^i\,, \\ d_W\,\bar\omega^i = 0\,. \end{cases}$$

Change the basis of Weil (G) in introducing new elements Ω^i given by

$$(8)\qquad \frac{1}{2}\Omega^i = \bar\omega^i + \frac{1}{2}c^i_{jk}\,\omega^j\omega^k\,.$$

Then, we can also write

$$(5)'\qquad \mathrm{Weil}\,(G) = \mathrm{Free}\,(\omega^i, \Omega^i)\,,\ \text{with}$$

$$(9)\quad \begin{cases} \deg\omega^i = 1\,, \qquad \deg\Omega^i = 2\,, \\ d_W\,\omega^i = -\frac{1}{2}\,c^i_{jk}\,\omega^j\omega^k + \frac{1}{2}\Omega^i\,, \\ d_W\,\Omega^i = c^i_{jk}\,\Omega^j\omega^k\,. \end{cases}$$

Now, for any $g \in G$ let us define a linear operator Ad_g acting on Weil (G), such that $Ad_g\,\omega^i$ is the same as in $A^L(G)$, and $Ad_g\,\bar\omega^i$ as in $Ad_g\,\omega^i$, with each ω^j replaced by $\bar\omega^j$. Similarly, we define for any $X \in g$, the derivation ad_X on Weil (G) which is the same as in $A^L(G)$ for $ad_X\,\omega^i$, and also the same for $ad_X\,\bar\omega^i$ with each ω^j replaced by $\bar\omega^j$.

Then:

$$(10)\quad \begin{cases} ad_k\,\omega^i = -\,c^i_{kj}\,\omega^j\,, \\ ad_k\,\bar\omega^i = -\,c^i_{kj}\,\bar\omega^j\,. \end{cases}$$

From these, it follows that in Weil (G)

$$(11)\qquad ad_k\,\Omega^i = -\,c^i_{kj}\,\Omega^j \quad \text{and}$$

$$(12)\qquad Ad_g\,d_W = d_W\,Ad_g \qquad \text{or}$$

$$(12)'\qquad ad_k\,d_W = d_W\,ad_k\,.$$

Lemma 1: For any $h \in H$, the linear subspaces of Weil (G) spanned by ω^a, by ω^α and by Ω^α are each invariant under Ad_h. Moreover, we have

$$(13)\quad \begin{cases} ad_\beta\,\omega^a = -\,c^a_{\beta b}\,\omega^b\,, \\ ad_\beta\,\omega^\alpha = -\,c^\alpha_{\beta\gamma}\,\omega^\gamma\,, \\ ad_\beta\,\Omega^\alpha = -\,c^\alpha_{\beta\gamma}\,\Omega^\gamma\,. \end{cases}$$

Proof: These follow directly from the facts that H is a subgroup so that $c^a_{\alpha\beta} = 0$, and G is compact so that c^k_{ij} is anti-symmetric in all three indices i, j, k.

Now, in $\mathrm{Weil}(G)$ let us consider the sub-algebra $\mathrm{Weil}^I(G)$ consisting of all elements ξ which are Ad_g-invariant or $Ad_g\,\xi = \xi$ for all $g \in G$ or $ad_k\,\xi = 0$ for all indices k. Since $ad_k\,d_W = d_W\,ad_k$, $\mathrm{Weil}^I(G)$ is in fact a sub-DGA of $\mathrm{Weil}(G)$.

Lemma 2: Both, $\mathrm{Weil}(G)$ and $\mathrm{Weil}^I(G)$ are homologically trivial.

Proof: The assertion about $\mathrm{Weil}(G)$ follows directly from its very definition. Let ξ be a cycle of $\mathrm{Weil}^I(G)$, then it is also a cycle of $\mathrm{Weil}(G)$, so that $\xi = d_W\,\eta$ for some $\eta \in \mathrm{Weil}(G)$. Since ξ is Ad_g-invariant for any $g \in G$ and G is compact, $\xi = \int_G Ad_g\,\xi dg = d_W \int_G Ad_g\,\eta dg = d_W\,\eta'$, where $\eta' = \int_G Ad_g\,\eta dg$, being Ad_g-invariant, $\in \mathrm{Weil}^I(G)$, and the lemma is also proved for $\mathrm{Weil}^I(G)$.

For latter use, we shall introduce a differential in $\mathrm{Weil}^I(G)$ as follows. For the differential d_W in $\mathrm{Weil}(G)$ let us rewrite it in the following form:

$$d_W\,\omega^i = -\frac{1}{2}c^i_{jk}\,\omega^j\omega^k + \frac{1}{2}\Omega^i = \frac{1}{2}c^i_{jk}\,\omega^j\omega^k + \frac{1}{2}\Omega^i + \omega^j\,ad_j\,\omega^i\ ,$$

$$d_W\,\Omega^i = c^i_{jk}\,\Omega^j\omega^k = -c^i_{jk}\,\omega^j\Omega^k = 0 + \omega^j\,ad_j\,\Omega^i\ .$$

If we introduce d_G and d_Ω by setting

$$d_G\,\omega^i = -\frac{1}{2}c^i_{jk}\,\omega^j\omega^k\ ,$$

$$d_\Omega\,\omega^i = \frac{1}{2}\Omega^i\ ,$$

with all the other differentials equal to 0, then we have

$$d_W = -d_G + d_\Omega + \omega^j\,ad_j\ .$$

Like in the beginning of this section about the differential in W^I_B, we see that in $\mathrm{Weil}^I(G)$ we can use $-d_G + d_\Omega$ instead of d_W for the determination of its homology. We also remark that d_G operating on $A^I(G) \subset \mathrm{Weil}^I(G)$ in the same way as the differential in $A^I(G)$ is considered a DGA by itself. Remark, however, that $A^I(G)$ is not a sub-DGA of $\mathrm{Weil}^T(G)$.

Now, we come to the pair of a compact Lie group G and its sub-group H, as before.

Definition: For the pair (G,H) we define Weil(G,H) as the sub-DGA of Weil(G) generated by $\omega^a, \omega^\alpha, \Omega^\alpha$ alone:

$$\mathrm{Weil}(G,H) = \mathrm{Free}(\omega^a, \omega^\alpha, \Omega^\alpha) = \mathrm{Free}(\omega^a) \otimes \mathrm{Weil}(H) .$$

We define the Weil algebra $\mathrm{Weil}^I(G,H)$ as the sub-DGA of Weil(G,H) consisting of forms invariant under Ad_h for all $h \in H$ or those ξ with

$$\mathrm{ad}_\alpha \xi = 0$$

for all indices α.

Lemma 3: The linear operator

$$d_{W^1} = -d_G + d_\Omega$$

will also induce a differential in $\mathrm{Weil}^I(G,H)$.

Proof: The d_{W^1} can be expressed in a more explicit form as follows:

$$d_{W^1}\, \omega^a = \frac{1}{2} c^a_{jk}\, \omega^j \omega^k ,$$

$$d_{W^1}\, \omega^\alpha = \frac{1}{2} c^\alpha_{jk}\, \omega^j \omega^k + \frac{1}{2} \Omega^\alpha ,$$

$$d_{W^1}\, \Omega^\alpha = 0 .$$

Only $d^2_{W^1} = 0$ requires proof. Now, by direct calculation we find

$$d^2_{W^1}\, \omega^i = -\frac{1}{2} \Omega^\beta\, \mathrm{ad}_\beta\, \omega^i ,$$

$$d^2_{W^1}\, \Omega^\alpha = 0 .$$

Since elements ξ in $\mathrm{Weil}^I(G,H)$ are such that $\mathrm{ad}_\beta \xi = 0$, we see, like in the beginning of this section, that $d^2_{W^1} = 0$ in Weil(G,H), as to be proved.

Henceforth, we shall understand $\mathrm{Weil}^I(G,H)$ to be a DGA with differential, induced by the above d_{W^1}. From the very definition we have then

Lemma 4: The module-morphism

$$\sigma : \mathrm{Weil}^I(G) \to \mathrm{Weil}^I(G,H)$$

obtained by sending all Ω^a to 0 is a DGA-morphism.

To go further, let us put the d_{W^1} in another form by using various properties of c^k_{ij} :

$$d_{W^1}\omega^a = \frac{1}{2}c^a_{jk}\,\omega^j\omega^k$$

$$= \frac{1}{2}c^a_{bc}\,\omega^b\omega^c + c^a_{\beta b}\,\omega^\beta\omega^b$$

$$= \frac{1}{2}c^a_{bc}\,\omega^b\omega^c - \omega^\beta\,\mathrm{ad}_\beta\,\omega^a ,$$

$$d_{W^1}\omega^\alpha = \frac{1}{2}c^\alpha_{jk}\,\omega^j\omega^k + \frac{1}{2}\Omega^a$$

$$= \frac{1}{2}c^\alpha_{bc}\,\omega^b\omega^c + \frac{1}{2}c^\alpha_{\beta\gamma}\,\omega^\beta\omega^\gamma + \frac{1}{2}\Omega^\alpha$$

$$= \frac{1}{2}c^\alpha_{bc}\,\omega^b\omega^c - \frac{1}{2}c^\alpha_{\beta\gamma}\,\omega^\beta\omega^\gamma + \frac{1}{2}\Omega^\alpha - \omega^\beta\,\mathrm{ad}_\beta\,\omega^\alpha ,$$

$$d_{W^1}\Omega^\alpha = 0$$

$$= -c^\alpha_{\beta\gamma}\,\omega^\beta\Omega^\gamma - \omega^\beta\,\mathrm{ad}_\beta\,\Omega^\alpha .$$

We have then

$$d_{W^1} = d_H + d_B - \omega^\beta\,\mathrm{ad}_\beta$$

in which d_H , d_B are given by

$$d_B\,\omega^a = \frac{1}{2}c^a_{bc}\,\omega^b\omega^c ,$$

$$d_B\,\omega^\alpha = \frac{1}{2}c^\alpha_{bc}\,\omega^b\omega^c ,$$

$$d_H\,\omega^\alpha = -\frac{1}{2}c^\alpha_{\beta\gamma}\,\omega^\beta\omega^\gamma + \frac{1}{2}\Omega^\alpha ,$$

$$d_H\,\Omega^\alpha = -c^\alpha_{\beta\gamma}\,\omega^\beta\Omega^\gamma$$

with other d_B and d_H equal to 0 . Like in the beginning of this section, we can then also replace d_{W^1} by

$$d_{W^2} = d_H + d_B$$

which gives the same final result as d_{W^1} in $\mathrm{Weil}^I(G,H)$. We also remark that in comparison with formulae (9), the d_H restricted on $\mathrm{Weil}(H) \subset \mathrm{Weil}^I(G,H)$ is just the same as d_W in Weil(H) .

Define now a module-morphism

$$j_B : W_B^I \to \mathrm{Weil}^I(G,H)$$

by

$$j_B(\omega^a) = -\omega^a$$

which is legitimate, since Ad_h-invariant elements will be sent to Ad_h-invariant elements for any $h \in H$. From the differentials d_B in W_B^I and d_{W^2} in $\mathrm{Weil}^I(G,H)$ we see that j_B is in fact a DGA-morphism.

The main object of this section is to prove the following

Theorem: The DGA-morphism j_B is an H-isomorphism.

It is clear that the theorem is a direct consequence of the following

Lemma 5: For any $\xi \in \mathrm{Weil}^I(G,H)$ with $d_{W^2}\xi \in \mathrm{Im}\, j_B$, we have

$$\xi = d_{W^2}\eta + \zeta$$

for some $\eta \in \mathrm{Weil}^I(G,H)$ and $\zeta \in \mathrm{Im}\, j_B$.

Proof: For any $\xi \in \mathrm{Weil}^I(G,H)$ we shall denote by $\deg_H \xi$ the maximum of total degree in ω^α and Ω^α of the terms of ξ. Now, we shall prove the lemma by induction on $\deg_H \xi$ which is trivial in case $\deg_H \xi = 0$, i.e., for $\xi \in \mathrm{Im}\, j_B$.

Consider thus $\xi \in \mathrm{Weil}^I(G,H)$ and suppose that the lemma has been proved for elements of $\deg_H$ < that of ξ. Write for this purpose

$$\xi = \xi_0 + \xi^1$$

with $\deg_H$ of each term in ξ_0, equal to $\deg_H \xi > \deg_H \xi^1$. As Ad_h clearly preserves $\deg_H$ and ξ is Ad_h-invariant so ξ_0 is also Ad_h-invariant for all $h \in H$ or $\xi_0 \in \mathrm{Weil}^I(G,H)$, cf. (13).

Now,

$$d_{W^2}\xi = d_H\xi_0 + (d_B\xi_0 + d_{W^2}\xi^1) \in \mathrm{Im}\, j_B$$

or $\deg_H d_{W^2}\xi = 0$ by hypothesis. Since $\deg_H$ of the element in the bracket is < that of $d_H\xi_0$ if the latter is not 0, we necessarily have $d_H\xi_0 = 0$.

Let

$$\xi_0 = \omega^{a_1}\dots\omega^{a_s}\xi_{a_1\dots a_s}$$

with $\xi_{a_1 \dots a_s} \in \mathrm{Free}(\omega^\alpha, \Omega^\alpha)$. Then we have

$$d_H \xi_0 = (-1)^s \omega^{a_1} \dots \omega^{a_s} d_H \xi_{a_1 \dots a_s} = 0$$

and therefore

$$d_H \xi_{a_1 \dots a_s} = 0 .$$

By what we have remarked, $\mathrm{Weil}(H) \subset \mathrm{Weil}^I(G,H)$ and is homologically trivial, so that

$$\xi_{a_1 \dots a_s} = d_H \eta^1_{a_1 \dots a_s}$$

for some $\eta^1_{a_1 \dots a_s} \in \mathrm{Weil}(H)$. Hence,

$$\xi_0 = d_H(\pm \omega^{a_1} \dots \omega^{a_s} \eta^1_{a_1 \dots a_s}) = d_H \eta^1, \quad \text{say.}$$

Since ξ_0 is Ad_h-invariant

$$\xi_0 = \int_H \mathrm{Ad}_h \xi_0 \, dh = d_H \int_H \mathrm{Ad}_h \eta^1_0 \, dh$$

$$= d_H \eta_0 ,$$

where

$$\eta_0 = \int_H \mathrm{Ad}_h \eta^1_0 \, dh$$

is Ad_h-invariant or $\eta_0 \in \mathrm{Weil}^I(G,H)$.

It follows that

$$\xi - d_{W^2} \eta_0 = \xi^1 - d_B \eta_0 \in \mathrm{Weil}^I(G,H)$$

too and has a $\deg_H <$ that of ξ. Moreover, $d_{W^2}(\xi - d_{W^2} \eta_0) = d_{W^2} \xi \in \mathrm{Im}\, j^I_B$. By induction hypothesis we have therefore

$$\xi - d_{W^2} \eta_0 = d_{W^2} \eta_1 + \zeta_1$$

for some $\eta_1 \in \mathrm{Weil}^I(G,H)$ and $\zeta_1 \in \mathrm{Im}\, j^I_B$.

Hence, $\xi = d_{W^2}(\eta_0 + \eta_1) + \zeta_1$ with $\eta_0 + \eta_1 \in \mathrm{Weil}^I(G,H)$ and $\zeta_1 \in \mathrm{Im}\, j^I_B$. This proves the lemma and, thus, also the theorem.

V.4 THE CARTAN ALGEBRA AND THE THEOREM OF CARTAN

In the preceding section, we have introduced the Weil algebra $\mathrm{Weil}^I(G,H)$ which is H-isomorphic to W_B^I under a DGA-morphism j_B. In this section, we shall study more in detail the structure of $\mathrm{Weil}^I(G,H)$, reducing this further to a sub-DGA called Cartan algebra to which it is again H-isomorphic under a certain DGA-morphism. For this purpose, we shall first introduce some important conceptions and state without proof the theorem of Hopf, et al as follows.

Definition: A form

$$p = p_{i_1 \dots i_s} \omega^{i_1} \dots \omega^{i_s} \in A^L(G) , \quad s \geq 1$$

is said to be primitive if

1°. $p \in A^I(G)$, or $\mathrm{ad}_i p = 0$ for all indices i .

2°. $\partial_{i_1} \dots \partial_{i_m} p$ is orthogonal to $A^I(G)$

for all m-tuples $(i_1, \dots, i_m)$ with $1 \leq m \leq s-1$.

Theorem of Hopf: The primitives of $A^L(G)$ span a finite dimensional linear space $P(G)$ with a basis

$$p_1 , p_2 , \dots , p_N$$

for which all p_i are of odd degree and

$$\deg p_1 \leq \deg p_2 \leq \dots \leq \deg p_N .$$

Moreover, $A^I(G)$ is an exterior algebra over these primitives as free generators

$$A^I(G) \approx \mathrm{Extr}\,(p_1 , \dots , p_N) .$$

Cor.: For a compact Lie group

$$H^*_{\underline{R}}(G) \approx \mathrm{Extr}\,(p_1 , \dots , p_N) .$$

Definition: In the Weil algebra of G:

$\mathrm{Weil}^I(G) \subset \mathrm{Weil}(G) = \mathrm{Free}\,(\omega^i, \Omega^i)$, an element ξ will be said to be of

$\deg_\omega \xi = k$ if the maximum of total degrees in ω^i of terms in ξ is equal to k. Write any such element ξ in the form

$$\xi = \xi_k + \xi_{k-1} + \dots + \xi_0$$

for which each ξ_i consists of terms of $\deg_\omega$ equal to i. Then, ξ_k will be called the Lead of ξ:

$$\xi_k = \text{Lead}\,\xi .$$

Definition: A form $\xi \in A^I(G) \subset \text{Weil}^I(G)$ is said to be transgressive if there is some form $\eta \in \text{Weil}^I(G)$, such that

$$\text{Lead}\,\eta = \xi$$

and

$$\deg_\omega d_W \eta = 0 ,$$

or $d_W\eta$ is generated by Ω^i alone. In that case $d_W\eta$ is called a transgression of ξ.

Notation: $S^I(G)$ = sub-algebra of $\text{Weil}^I(G)$ consisting of forms generated by Ω^i alone, or of forms with $\deg_\omega$ equal to 0.

Theorem of Cartan-Chevalley-Weil:

Consider $A^I(G)$ as a sub-GA of $\text{Weil}^I(G)$. Then, a form in $A^I(G)$ is transgressive if and only if it is primitive. Moreover, for each primitive, its transgressions are unique up to decompasables. If we choose for each primitive p_μ a fixed transgression $s_\mu \in S^I(G)$ so that

$$(1)\quad \begin{cases} p_\mu = \text{Lead}\, q_\mu , \\ d_W q_\mu = s_\mu \end{cases}$$

for some $q_\mu \in \text{Weil}^I(G)$, then $S^I(G)$ is a polynomial algebra on s_μ as free generators.

Consider now the Weil-algebra associated to the pair (G,H), viz.

$$\text{Weil}^I(G,H) \subset \text{Weil}\,(G,H) = \text{Free}\,(\omega^i, \Omega^\alpha) .$$

As already shown in Lemma 4 of § V.3, the morphism

$$\sigma : \mathrm{Weil}^I(G) \to \mathrm{Weil}^I(G,H)$$

obtained by sending Ω^a to 0 while keeping the other ω^i and Ω^α will induce a DGA-morphism. Moreover, it is clear that

$$\sigma(S^I(G)) \subset S^I(H) ,$$

and σ is identity on $A^I(G)$ which is a sub-GA (but not sub-DGA) of both $\mathrm{Weil}^I(G)$ and $\mathrm{Weil}^I(G,H)$.

Now, for the compact Lie group H $A^I(H)$ is also an exterior algebra with some primitives (degrees all odd in non-decreasing order)

$$p_{H1} , p_{H2} , \dots , p_{HM}$$

as free generators. To each such primitive $p_{H\lambda}$ we may choose a fixed transgression $s_{H\lambda}$ such that

$$(2) \quad \begin{cases} p_{H\lambda} = \mathrm{Lead}\, q_{H\lambda} , \\ d_{W,H}\, q_{H\lambda} = s_{H\lambda} \end{cases}$$

for some $q_{H\lambda} \in \mathrm{Weil}^I(H)$, where $d_{W,H}$ is the differential in $\mathrm{Weil}^I(H)$ as d_W in $\mathrm{Weil}^I(G)$. Since $\sigma(S^I(G)) \subset S^I(H)$ and $S^I(G)$, $S^I(H)$ are polynomial algebras with s_μ and $s_{H\lambda}$ respectively as free generators, we have

$$\sigma s_\mu = P_\mu(s_{H\lambda})$$

for some polynomials $_\mu$ in $s_{H\lambda}$.

<u>Definition</u>: The <u>Cartan algebra</u> C associated to the pair (G,H) is the graded algebra

$$C = \mathrm{Polym}\,(s_{H\lambda}) \otimes \mathrm{Extr}\,(p_\mu)$$

with a differential d_C given by

$$(3) \quad \begin{cases} d_C\, p_\mu = \sigma s_\mu = P_\mu(s_{H\lambda}) , \\ d_C\, s_{H\lambda} = 0 . \end{cases}$$

We come now to the main theorem of this section, which is due to H. Cartan.

Theorem: Let the transgressions s_μ be determined from the primitives p_μ as in (1) . Then, there is a DGA-morphism

$$j_C : C \to \mathrm{Weil}^I(G,H)$$

defined by

$$(4) \quad \begin{cases} j_C\, p_\mu = \sigma q_\mu \ , \\ j_C\, s_{H\lambda} = s_{H\lambda} \end{cases}$$

which is also an H-isomorphism.

Proof: Remark first that j_C is a monomorphism since the lead of each q_μ is just p_μ . That j_C induces a DGA-morphism is quite clear from the very definition. That j_C is an H-isomorphism follows clearly from the following

Lemma: For any $\xi \in \mathrm{Weil}^I(G,H)$ with $d_{W^I}\xi \in \mathrm{Im}\, j_C$, we have

$$\xi = d_{W^I}\eta + \zeta$$

for some $\eta \in \mathrm{Weil}^I(G,H)$ and $\zeta \in \mathrm{Im}\, j_C$.

Proof: Let us use the expression

$$d_{W^I} = -d_G + d_\Omega$$

for the differential in $\mathrm{Weil}^I(G,H)$ as given in Lemma 3 of § V.3 . We remark also that d_G on $A^I(G) \subset \mathrm{Weil}^I(G,H)$ is just the same differential in $A^I(G)$ considered as a DGA by itself.

We shall prove now the Lemma by induction ond $\deg_\omega \xi$. For $\deg_\omega \xi$ equal to 0 there is nothing to prove. Consider therefore an element $\xi \in \mathrm{Weil}^I(G,H)$ with $\deg_\omega \xi = k > 0$, assuming that the lemma is already proved for elements of $\deg_\omega$ less than k .

Write now ξ in the form

$$\xi = \xi_0 + \xi'$$

with

$$\xi_0 = \mathrm{Lead}\, \xi \ , \quad \deg_\omega \xi' < k \ .$$

It follows that

$$d_{W^1}\xi = -d_G\xi_0 + \xi''$$

with

$$\xi'' = d_\Omega\xi_0 + d_{W^1}\xi'$$

of $\deg_\omega$ less than $\deg_\omega d_G\xi_0$ if $d_G\xi_0$ is not 0.

It is clear that Ad_h preserves $\deg_\omega$ so that ξ_0 is Ad_h-invariant as ξ for any $h \in H$. Let

$$\xi_0 = s_\tau \xi_{0\tau}$$

with $s_\tau \in \text{Polym}(\Omega^\alpha)$ and $\xi_{0\tau} \in \text{Extr}(\omega^i)$ where τ runs over some index set. Furthermore, $d_{W^1}\xi \in \text{Im}\, j_C$ by hypothesis would imply that $d_G\xi_0$ is of the form

$$d_G\xi_0 = s_\tau d_G\xi_{0\tau} = s_{H\nu}\pi_\nu$$

with

$$s_{H\nu} \in \text{Polym}(\Omega^\alpha), \quad \pi_\nu \in \text{Extr}(p_\mu)$$

and ν running over some index sets.

Comparing the right hand sides we see that for each ν, $\pi_\nu = d_G\xi_{0\tau}$ for some τ, so that $\pi_\nu \sim 0$ in $A^I(G)$, and hence by Theorem of Hopf each π_ν is itself 0. We have then $d_G\xi_{0\tau} = 0$ for each τ and, again by Theorem of Hopf, we have

$$\xi_{0\tau} = d_G\eta_{0\tau} + \pi'_\tau$$

for some $\eta_{0\tau} \in \text{Extr}(\omega^i)$ and $\pi'_\tau \in \text{Extr}(p_\mu)$.

Consequently

$$\xi_0 = d_G(s_\tau\eta_{0\tau}) + s_\tau\pi'_\tau .$$

Since ξ_0 is Ad_H-invariant we get then

$$\xi_0 = d_G\eta_0 + \zeta_0 + \zeta'_0 ,$$

where

$$\eta_0 = \int_H Ad_h(s_\tau\eta_{0\tau})\,dh \in \text{Weil}^I(G,H) ,$$

$$\zeta_0 = \int_H (Ad_h s_\tau)\,dh \cdot \rho_\tau \in \text{Im}\, j_C ,$$

$$\zeta'_0 = \int_H (Ad_h s_\tau)\,dh \cdot (\pi'_\tau - \rho_\tau) ,$$

in which ρ_τ is obtained from π'_τ in replacing each p_μ by σq_μ, and

$\deg_\omega \zeta_0' < k$.

Now, $\xi + d_{W^1}\eta_0 - \zeta_0 = d_\Omega \eta_0 + \xi' + \zeta_0'$ has a $\deg_\omega$ less than k and clearly

$$d_{W^1}(\xi + d_{W^1}\eta_0 - \zeta_0) \in \operatorname{Im} j_C$$

so that by induction hypothesis

$$\xi + d_{W^1}\eta_0 - \zeta_0 = d_{W^1}\eta_1 + \zeta_1$$

with $\eta_1 \in \mathrm{Weil}^I(G,H)$ and $\zeta_1 \in \operatorname{Im} j_C$.

Then,

$$\xi = d_{W^1}(\eta_1 - \eta_0) + \zeta_0 + \zeta_1$$

with $\eta_1 - \eta_0 \in \mathrm{Weil}^I(G,H)$ and $\zeta_0 + \zeta_1 \in \operatorname{Im} j_C$.
This completes the induction and the Lemma as well as the above Theorem of Cartan are proved.

In combining together the above theorem and the results in preceding sections we have a diagram below of DGA-morphisms which are all H-isomorphisms:

$$\begin{array}{ccccc} C & \xrightarrow{j_C} & \mathrm{Weil}^I(G,H) & \xleftarrow{j_B} & W_B^I \\ \uparrow{\scriptstyle \rho_C} & & & & \uparrow{\scriptstyle \pi^*} \\ I^*(M) & \xrightarrow{\rho_M} & A^*(M) & \xleftarrow{\text{incl.}} & A^L(M) \end{array}$$

By Chap. II, there will then exist a DGA-morphism $\rho_C : I^*(M) \to C$ which is an H-isomorphism too so that we get the following main theorem of this chapter:

<u>Theorem of H. Cartan:</u> For the homogeneous space $M = G/H$ with G a compact connected Lie group and H a closed connected subgroup of G we have

$$I^*(M) \approx \min C \quad \text{and} \quad H^*_{\mathbb{R}}(M) \approx H(C)$$

where C is the Cartan algebra associated to the pair (G,H) .

We add now an alternative treatment of the Cartan theorem, as follows. To each compact connected Lie group G there is a <u>classifying space</u> B_G which is the base space of an universal fibration

$$(\mathcal{G}) \qquad G \subset E_G \xrightarrow{g} B_G$$

with fiber G and contractible fiber space E_G. Now, for G we have $I^*(G) \approx H^*(G)$, so we may identify these two and we have ($H^* = H^*_{\underline{R}}$, below)

$$H^*(G) = I^*(G) \approx \mathrm{Extr}\,(p_1, \ldots, p_N)$$

where p_μ are the primitive elements spanning a linear space P_G. It is known that all p_μ are transgressive elements and $H^*(B_G) \approx I^*(B_G)$ are polynomial algebras generated by the transgressions of p_μ. In more details, let us identify $H^*(B_G) = I^*(B_G)$ and denote their quotient with respect to the subalgebra D_G of their decomposables by Q_G. Then, there is some isomorphism

$$\lambda_G : P_G \approx Q_G .$$

Denote by $p_G : H^*(B_G) \to Q_G$ the natural projection and take for each p_μ an element $s_\mu \in H^*(B_G)$ such that $p_G s_\mu = \lambda_G p_\mu$.
Then, $\tau_G : P_G \to H^*(B_G)$ defined by $\tau_G(p_\mu) = s_\mu$ is a transgression and

$$H^*(B_G) \approx I^*(B_G) \approx \mathrm{Polym}\,(s_1, \ldots, s_N) .$$

Similarly, for the closed connected subgroup H of G, we shall have $H^*(H) \approx I^*(H) \approx \mathrm{Extr}\,(p_{H_1}, \ldots, p_{HL})$ where $p_{H\lambda}$ are primitives spanning a linear subspace P_H, etc. so that we have a commutative diagram below:

$$\begin{array}{ccccccc}
\mathrm{Extr}\,(p_\mu) \approx I^*(B_G) \supset & P_G & \xrightarrow[\approx]{\lambda_G} & Q_G & \xleftarrow{p_G} & I^*(B_G) & \approx \mathrm{Polym}\,(s_\mu) \\
\downarrow i^I & \downarrow i^I & & \downarrow g^I & & \downarrow g^I & \\
\mathrm{Extr}\,(p_{H\lambda}) \approx I^*(H) \supset & P_H & \xrightarrow[\lambda_H]{\approx} & Q_H & \longleftarrow & I^*(B_H) & \approx \mathrm{Polym}\,(s_{H\lambda})
\end{array}$$

In the diagram $i : H \subset G$ is the inclusion and $i^I(P_G)$ or $i^H(P_G) \subset P_H$ follows from a theorem of Samalson. The morphism $g^I : I^*(B_G) \to I^*(B_H)$ or $g^H : H^*(B_G) \to H^*(B_H)$ is induced by the projection g in some natural fibration

$$G/H \subset B_H \xrightarrow[g]{} B_G .$$

In fact, the elements s_μ and $s_{H\lambda}$ may be identified to those of the Weil algebra $\mathrm{Weil}^I(G,H)$ in § V.3, 4. Hence, the Cartan algebra C associated to the homogeneous space $M = G/H$ may also be put in the following form:

$$C = H^*(B_H) \otimes H^*(G)$$

or

$$C = I^*(B_H) \otimes I^*(G)$$

$$\approx \text{Polym}(s_{H\lambda}) \otimes \text{Extr}(p_\mu) ,$$

with differential d_C given by

$$d_C p_\mu = g^I \tau_G p_\mu = g^I s_\mu ,$$

$$d_C s_\mu = 0 .$$

Chapter VI

EFFECTIVE COMPUTATION AND AXIOMATIC SYSTEM OF I^*-MEASURE

VI.1 AN EXTENSION THEOREM

The aim of this section is to prove the following theorem, which is at the basis of many considerations about I^*-measures, in particular for their calculability with respect to union- and cone-constructions, as described in later sections of this Chapter. The usefulness may also be seen from the Proposition of §II.5.

Extension Theorem: Let K be any connected countable simplicial complex in weak topology and L a connected subcomplex with inclusion $i : L \subset K$. Then, the DGA-morphism

$$i^A : A^*(K) \to A^*(L)$$

of the deRham-Sullivan algebras is onto.

To prove the theorem let us first make some preparations. Let Δ^n be an n-simplex with vertices $v_0, v_1, \ldots, v_n$.
Denote by $\dot{\Delta}^n$ the boundary of Δ^n. Let v be a new vertex and $v\Delta^n, v\dot{\Delta}^n$ be the complex obtained from Δ^n and $\dot{\Delta}^n$ by adjoining such a new vertex, i.e., the cone erected over Δ^n and $\dot{\Delta}^n$ with vertex at v. Then, the Extension Theorem will follow clearly from the following two lemmas:

Extension Lemma 1: The DGA-morphism

$$A^*(v\Delta^n) \to A^*(v\dot{\Delta}^n)$$

induced by the inclusion $v\dot{\Delta}^n \subset v\Delta^r$ is onto.

Extension Lemma 2: The DGA-morphism $A^*(v\dot{\Delta}^n) \to A^*(\dot{\Delta}^n)$ induced by the inclusion $\dot{\Delta}^n \subset v\dot{\Delta}^n$ is onto.

For the proofs we shall first introduce some notations. We shall denote by N the index set

$$N = \{0, 1, \ldots, n\} .$$

For any subset I of N the complementary set in N will then be denoted by $N-I$.

For any non-empty sub-set $I = (i_0, i_1, \dots, i_m)$ of N the simplex spanned by the vertices $v_{i_0}, v_{i_1}, \dots, v_{i_m}$ will be denoted by Δ_I which is of dimension $|I| - 1$, where $|I|$ is the number of indices in the set I. In particular, for any $k \in N$, $\Delta_{\{k\}}$ will be denoted simply by Δ_k, similarly for others.

For any non-empty subset $I \subset N$ let us define a projection

$$\pi_I : \Delta^n \to v\Delta_{N-I}$$

as the linear map determined by:

$$\pi_I(v_i) = v, \qquad i \in I,$$

$$\pi_I(v_j) = v_j, \qquad j \in N-I,$$

$$\pi_I(v) = v.$$

We also denote by i_I the inclusion

$$i_I : v\Delta_{N-I} \subset v\Delta^n .$$

<u>Proof of Lemma 1</u>: Consider an arbitrary element $w \in A^*(v\dot{\Delta}^n)$ which is a compatible collection of differential forms on each subsimplex of $v\dot{\Delta}^n$. For any non-empty subset $I \subset N$, let the differential form on $v\Delta_{N-I}$ in this collection be $\check{w}_I$. Define now a differential form $\check{w}$ on the simplex $v\Delta^n$ by setting

$$\check{w} = \Sigma\, (-1)^{|I|+1} \cdot \hat{\pi}_I \check{w}_I ,$$

the summation being over all non-empty subsets I of N. Then clearly it is sufficient to prove that for any index $k \in N$, the restriction of $\check{w}$ on each simplex $v\Delta_{N-\{k\}}$ is $\check{w}_k = \check{w}_{\{k\}}$.

For this purpose let us write $\check{w}$ for each given $k \in N$ in the form

$$\check{w} = \hat{\pi}_k \check{w}_k + \Sigma'\, (-1)^{|I|+1} \cdot [\hat{\pi}_I \check{w}_I - \hat{\pi}_{I\cup\{k\}} \check{w}_{I\cup\{k\}}]$$

in which Σ' runs over all non-empty subsets I of N not containing k.

Consider now the following diagram of inclusions and projections in which the part with solid lines is clearly commutative:

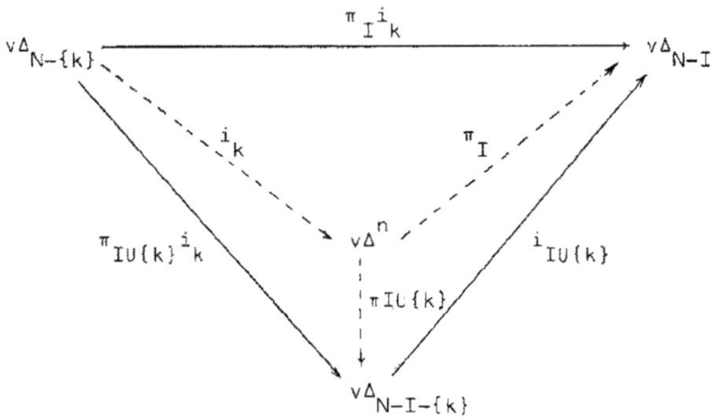

For the restrictions of forms we have

$$\hat{\pi}_I \check{w}_I \mid v\Delta_{N-\{k\}} = \hat{i}_k \hat{\pi}_I \check{w}_I = (\pi_I i_k)^\wedge \check{w}_I ,$$

$$\begin{aligned}\hat{\pi}_{I\cup\{k\}} \check{w}_{I\cup\{k\}} \mid v\Delta_{N-\{k\}} &= \hat{i}_k \hat{\pi}_{I\cup\{k\}} \check{w}_{I\cup\{k\}} \\ &= \hat{i}_k \hat{\pi}_{I\cup\{k\}} \hat{i}_{I\cup\{k\}} \check{w}_I \\ &= (i_{I\cup\{k\}} \pi_{I\cup\{k\}} i_k)^\wedge \check{w}_I .\end{aligned}$$

By the above-mentioned commutativity of the diagram we have therefore

$$\hat{\pi}_I \check{w}_I - \hat{\pi}_{I\cup\{k\}} \check{w}_{I\cup\{k\}} \mid v\Delta_{N-\{k\}} = 0 .$$

It follows that

$$\begin{aligned}\check{w} \mid v\Delta_{N-\{k\}} &= \hat{\pi}_k \check{w}_k \mid v\Delta_{N-\{k\}} \\ &= \hat{i}_k \hat{\pi}_k \check{w}_k = (\pi_k i_k)^\wedge \check{w}_k \\ &= \check{w}_k ,\end{aligned}$$

since $\pi_k i_k$ is identity on $v\Delta_{N-\{k\}}$.

Thus, the lemma is proved.

Proof of Lemma 2: Let the barycentric coordinate functions in the simplex $v\Delta^n$ corresponding to the vertices v, v_i $(0 \leq i \leq n)$ be respectively t and t_i. We shall put

$$T = t_0 + t_1 + \dots + t_n , \qquad T_k = T - t_k$$

$$T^{\vee} = t + T , \qquad T_k^{\vee} = t + T_k ,$$

$$dT = dt_0 + dt_1 + \dots + dt_n ,$$

$$dT_k = dT - dt_k ,$$

$$dT^{\vee} = dt + dT , \qquad dT_k^{\vee} = dt + dT_k .$$

Consider now any differential form w in $A^*(\dot{\Delta}^n)$ which is a compatible collection of forms on each simplex of $\dot{\Delta}^n$. The form on $\Delta_{N-\{k\}}$ in the collection will be, as before, denoted by w_k.

Let the degree of w be m. Then, we can take a formal expression of w_k in the form of

$$w_k^F = \Sigma\, g_{i_1 \dots i_m} dt_{i_1} \dots dt_{i_m} ,$$

in which the summation is over m-tuple of indices $(i_1 , \dots , i_m)$ none equal to k and $g_{i_1 \dots i_m}$ are polynomials in $t_0 , \dots , \hat{t}_k , \dots , t_n$. Note that w_k^F is not unique but is only determined up to multipliers of $T_k - 1$ and dT_k.

Let the maximum of the degrees of the polynomials $g_{i_1 \dots i_m}$ in $t_0 , \dots , \hat{t}_k \dots , t_m$ be n_k, and let $M > \operatorname{Max}_k (n_k + 2m)$. Let $\tilde{g}_{i_1 \dots i_m}$ be the polynomial obtained from $g_{i_1 \dots i_m}$ by replacing each t_i in it by $\frac{t_i}{1-t}$, clearing fractions, and then multiplying by $(1-t)^{n_k}$. We define now a formal form in $v\Delta_{N-\{k\}}$ by

$$\tilde{w}_k^F = (1-t)^{M-n_k-2m} \cdot \Sigma \tilde{g}_{i_1 \dots i_m} ((1-t)dt_{i_1} + t_{i_1} dt) \dots ((1-t)dt_{i_m} + t_{i_m} dt) .$$

First, let us show that $\tilde{w}_k^F$ is well-defined on $v\Delta_{N-\{k\}}$ as far as M is chosen sufficiently large. In fact, let w'^F_k be another formal expression of w_k. Then

$$w'^F_k - w_k^F = \xi_k^F \cdot dT_k + (T_k - 1)\eta_k^F$$

for certain forms ξ_k^F, η_k^F. In forming $\tilde{w}'^F_k - \tilde{w}_k^F$ we have to replace dT_k by

$$\sum_{i \neq k} [(1-t)dt_i + t_i\, dt] = (1-t)dT_k^{\vee} + (T_k^{\vee} - 1)dt$$

and $T_k - 1$ by

$$\sum_{i \neq k} t_i - (1-t) = T_k^{\vee} - 1$$

multiplied eventually by certain powers of $1-t$. It follows that $\tilde{w}'^F_k$ and $\tilde{w}^F_k$ differ by multiples of dT^v_K and T^v_k-1, so that $\tilde{w}^F_k$ determines a well-defined form $\tilde{w}_k$ on $v\Delta_{N-\{k\}}$ from w_k, independent of the chosen form w^F_k.

Now, the restriction of $\tilde{w}^F_k$ to $\Delta_{N-\{k\}}$ is obtained by setting $t=0$, $dt=0$, and thus clearly is w^F_k. As the forms w_k, $k \in N$ are compatible, so are also $\tilde{w}_k$ as far as M is sufficiently large. Thus, $\{\tilde{w}_k\}$ will determine a differential form in $A^*(v\dot{\Delta}^n)$ which extends the given form w in $A^*(\dot{\Delta}^n)$. This proves Lemma 2.

Remark: We have restricted the whole theory of DGA to the case that the basic field $\underline{k}$ is of characteristic 0. In fact, if $\underline{k}$ is of a non-zero characteristic p, then the Extension Theorem will no more be true, whatever the DGA corresponding to the deRham-Sullivan algebra, supposed to verify the deRham-Sullivan theorem, and some reasonable properties, may be. This may be seen from the following counter example, due to A. Borel, Cf. Coh. des espaces loc. compacts d'apres Leray, Lecture Notes in Math. No. 2 (1957):

Suppose that a certain DGA $A^*_p(K)$ on $\underline{Z}_p$ exists for any complex K, such that

(1) $A^*_p(K) \xrightarrow{i^A} A^*_p(L) \to 0$ is exact for the inclusion $i: L \subset K$ of a subcomplex L in a complex K.

(2) $H^*_{\underline{Z}_p}(K) \approx H(A^*_p(K))$ is an algebraic isomorphism.

Assertion: $X^p = 0$ for any $X \in H^{2s}_{\underline{Z}_p}(K)$, $s > 0$.

In fact, consider X as an element in $H_{2s}(A^*_p(K))$ by (2) and take $x \in X$. By (1) there will be some $x' \in A^{2s}_p(vK)$ such that $i^A x' = x$, where vK is the cone over K from a new vertex v, and $i: K \subset vK$ is the inclusion. Now, $dx'^p = px'^{p-1} = 0$, so that $x'^p = da'$ for some $a' \in A^*_p(vK)$, the complex vK being contractible. Then,

$$x^p = i^A x'^p = i^A da' = d(i^A a') \sim 0 \quad \text{in } K$$

so that $X^p = 0$ in $H(A^*_p(K)) \approx H^*_{\underline{Z}_p}(K)$.
This proves the assertion.

Now, take K to be the complex projective space CP_n with $n \geq p$ and $X \in H^2_{\underline{Z}_p}(CP_n)$ the generator of the ring $H^*_{\underline{Z}_p}(CP_n)$. Then $X^p \neq 0$ in $H^*_{\underline{Z}_p}(CP_n)$ which leads to a contradiction to the assertion.

VI.2 UNION OF COMPLEXES ALONG A COMMON SUBCOMPLEX

Let K', K'' be two connected simplicial complexes with a connected subcomplex $L = K' \cap K''$ in common. For the deRham-Sullivan's A^*-measures we have then a diagram

$$(1) \qquad A^*(K') \xrightarrow{i'^A} A^*(L) \xleftarrow{i''^A} A^*(K'')$$

in which the DGA-morphisms i'^A, i''^A induced by the inclusions are both onto by § VI.1 . Let K be the union of K' and K'' along the common subcomplex L to be denoted as

$$(2) \qquad K = K' \underset{L}{\cup} K'' .$$

Then, any element $\alpha \in A^*(K)$ can be identified as a pair of elements (α', α''), with $\alpha' \in A^*(K')$, $\alpha'' \in A^*(K'')$ the restrictions of α to K', K'' such that

$$i'^A \alpha' = i''^A \alpha'' \in A^*(L) .$$

This leads to the

Definition 1: For a pair of DGA's $\mathcal{A}'$, $\mathcal{A}''$ over a basic field $\underline{k}$ of characteristic 0 we define a DGA $\mathcal{A} = \mathcal{A}' \oplus \mathcal{A}''$ to be the one with

$$\mathcal{A}_p = \mathcal{A}'_p \oplus \mathcal{A}''_p$$

for each $p \geq 0$. The algebraic operations in $\mathcal{A}$ are defined by

$$(a'_1, a''_1) + (a'_2, a''_2) = (a'_1 + a'_2, a''_1 + a''_2) ,$$

$$(a'_1, a''_1) \cdot (a'_2, a''_2) = (a'_1 a'_2, a''_1 a''_2) ,$$

$$k(a', a'') = (ka', ka'') , \qquad k \in \underline{k} .$$

The differential d in $\mathcal{A}$ is defined by

$$d(a', a'') = (da', da'') .$$

It is easily verified that $\mathcal{A}$ is really a DGA over the same basic field $\underline{k}$ and will be called the direct sum of the DGA's $\mathcal{A}'$ and $\mathcal{A}''$.

Definition 2: Let $\mathcal{A}'$, $\mathcal{A}''$, $\mathcal{B}$ be a triple of DGA's with connecting DGA-morphisms f', f'' as in the diagram below:

$$\mathcal{A}' \xrightarrow{f'} \mathcal{B} \xleftarrow{f''} \mathcal{A}'' .$$

Let $\mathcal{A} = \mathcal{A}' \oplus \mathcal{A}''$ be the direct sum of $\mathcal{A}'$ and $\mathcal{A}''$. For each $p \geq 0$ let $\mathcal{C}_p$ be the sub-module of $\mathcal{A}_p$ consisting of all elements (a', a'') with $a' \in \mathcal{A}'_p$, $a'' \in \mathcal{A}''_p$ such that

$$f' a' = f'' a'' .$$

Then, the direct sum of all these $\mathcal{C}_p$ with algebraic operations and differentiations inherited from $\mathcal{A}$ will generate a sub-DGA $\mathcal{C}$ of $\mathcal{A}$ which will be called the <u>union</u> of $\mathcal{A}', \mathcal{A}''$ <u>along</u> $\mathcal{B}$ <u>under</u> f', f'' and will be denoted as

$$\mathcal{C} = \cup (f', f'') .$$

From these definitions the following proposition is quite clear:

<u>Proposition 1</u>: The A^*-measure is calculable with respect to the union construction of two complexes along a common subcomplex. More precisely, let K be the union of two complexes K', K'' along a common subcomplex L, then $A^*(K)$ is completely determined by the $A^*(K')$, $A^*(K'')$, $A^*(L)$ as well as the DGA-morphisms induced by the natural inclusions, viz. from the diagram (1). The explicit determination is given by

$$A^*(K) = \cup (i'^A, i''^A) .$$

Now, according to what we have explained in § I.3, the A^*-measure, though calculable with respect to certain geometrical constructions including the union along a common part, cannot be said to be a satisfactory one in that it is transfinite and non-invariant in character. For this reason, we shall replace it by I^*-measure which will meet all the three requirements as stated in § I.3. For this purpose, let us make first some preparations.

<u>Proposition 2</u>: Let (D) below be a <u>commutative</u> diagram of DGA's and DGA-morphisms:

$$\text{(D)} \qquad \begin{array}{ccccc} \mathcal{A}' & \xrightarrow{f'} & \mathcal{B} & \xleftarrow{f''} & \mathcal{A}'' \\ {\scriptstyle g'}\uparrow & & \uparrow{\scriptstyle g} & & \uparrow{\scriptstyle g''} \\ \mathcal{M}' & \xrightarrow{\varphi'} & \mathcal{N} & \xleftarrow{\varphi''} & \mathcal{M}'' \end{array}$$

Then, the pair $\tilde{g} = (g', g'')$ will define naturally a DGA-morphism

$$\tilde{g} : U(\varphi', \varphi'') \to U(f', f'')$$

by

$$\tilde{g}(m', m'') = (g'm', g''m''), \quad m' \in \mathcal{M}', \quad m'' \in \mathcal{M}''.$$

Suppose that in the diagram (D) the DGA-morphisms f', f'', φ' and φ'' are all onto, and the DGA-morphisms g', g'', g are all H-isomorphisms. Then $\tilde{g}$ is also an H-isomorphism.

<u>Proof</u>: Consider $\mathcal{A}', \mathcal{A}''$ all as chain complexes in neglecting the multiplicative structures. Then we have a commutative diagram (D') of chain complexes and chain maps, as shown below, for which the two horizontal lines both are exact sequences:

$$\text{(D')} \qquad \begin{array}{ccccccccc} 0 & \longrightarrow & U(f', f'') & \longrightarrow & \mathcal{A}' \oplus \mathcal{A}'' & \xrightarrow{\alpha} & \mathcal{B} & \longrightarrow & 0 \\ & & \uparrow{\scriptstyle \tilde{g}} & & \uparrow{\scriptstyle (g', g'')} & & \uparrow{\scriptstyle g} & & \\ 0 & \longrightarrow & U(\varphi', \varphi'') & \longrightarrow & \mathcal{M}' \oplus \mathcal{M}'' & \xrightarrow{\mu} & \mathcal{N} & \longrightarrow & 0 \end{array}$$

In the diagram, the morphism α is defined by $\alpha(a', a'') = f'a' - f''a''$, similarly for μ. This diagram will induce a commutative diagram (D''), as shown below, between the various H-group measures and the connecting group-morphisms by the classical algebraic-topological procedure

$$\text{(D'')} \qquad \begin{array}{ccccccccccc} \cdots \to & H_{q-1}(\mathcal{A}' \oplus \mathcal{A}'') & \to & H_{q-1}(\mathcal{B}) & \to & H_q(U(f', f'')) & \to & H_q(\mathcal{A}' \oplus \mathcal{A}'') & \to & H_q(\mathcal{B}) & \to \cdots \\ & \uparrow{\scriptstyle (g', g'')_H} & & \uparrow{\scriptstyle g_H} & & \uparrow{\scriptstyle \tilde{g}_H} & & \uparrow{\scriptstyle (g', g'')_H} & & \uparrow{\scriptstyle g_H} & \\ \cdots \to & H_{q-1}(\mathcal{M}' \oplus \mathcal{M}'') & \to & H_{q-1}(\mathcal{N}) & \to & H_q(U(\varphi', \varphi'')) & \to & H_q(\mathcal{M}' \oplus \mathcal{M}'') & \to & H_q(\mathcal{N}) & \to \cdots \end{array}$$

By hypothesis the vertical arrows, except the middle one, are all group-isomorphisms. By the five lemma this will also be true for the vertical arrow in the middle, so that $\tilde{g}$ induces an H-group-isomorphism. As $\tilde{g}$ is clearly a morphism as graded algebras, so g is an H-isomorphism of DGA's. This proves the proposition.

<u>Remark</u>: I owe the above proof to Prof. L. Taylor of the University of Notre Damo. The originial proof is elementary and direct, with no use of these machineries of classical algebraic topology.

<u>Proposition 3</u>: For DGA's $\mathcal{A}', \mathcal{A}'', \mathcal{B}$ and homotopic DGA-morphisms

$$f_0' \simeq f_1' : \mathcal{A}' \to \mathcal{B}$$

and

$$f_0'' \simeq f_1'' : \mathcal{A}'' \to \mathcal{B}$$

with f_0' , f_1' , f_0'' , f_1'' all onto we have

$$\min U(f_0', f_1'') \approx \min U(f_1', f_1'') .$$

Proof: In the notations of § II.2 $f_0' \simeq f_1'$ implies the existence of a commutative diagram of DGA's and DGA-morphisms as shown below:

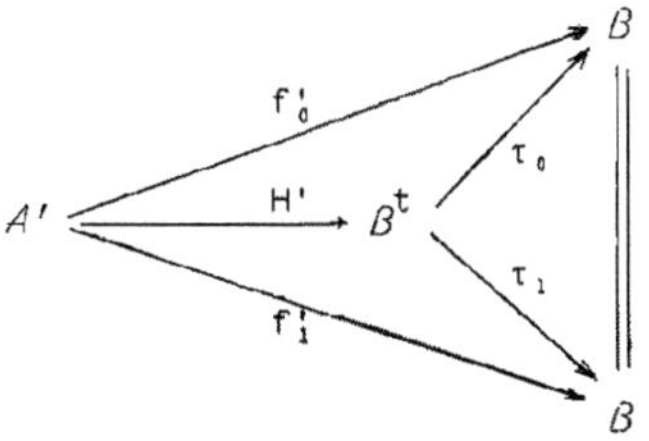

The morphism H' is in general not onto. However, let us form $\mathcal{A}'^t$ and extend H' to a DGA-morphism H'^t which sends t and dt in $\mathcal{A}'^t$ to t and dt in $\mathcal{B}^t$. Then the above diagram can be extended to the following commutative one with H'^t onto:

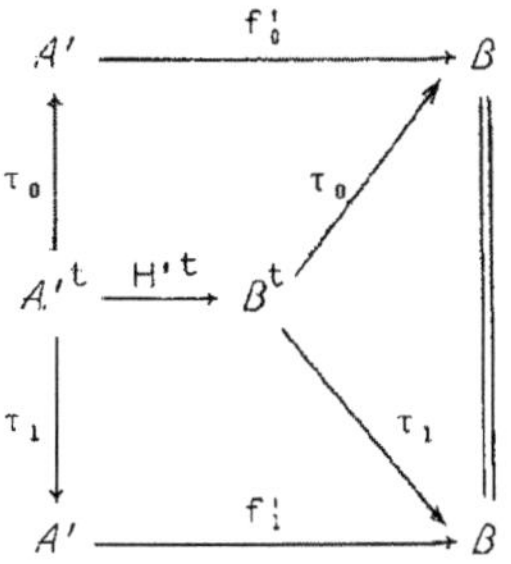

Similarly, for the case $f_0'' \simeq f_1''$. It follows that we have a commutative diagram

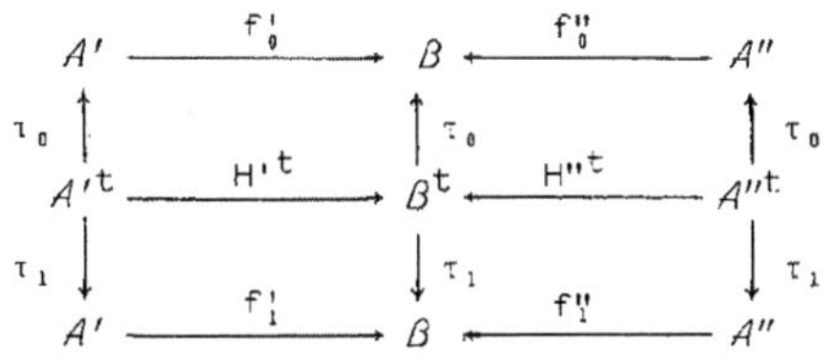

In the above diagram all the horizontal DGA-morphisms are onto and all the vertical ones are H-isomorphisms. By Prop. 2 the DGA-morphisms

$$\tilde{\tau}_0 = (\tau_0 , \tau_0) : U(H'^t, H''^t) \to U(f_0' , f_0'')$$

and

$$\tilde{\tau}_1 = (\tau_1 , \tau_1) : U(H'^t, H''^t) \to U(f_1' , f_1'')$$

both are H-isomorphisms. Hence,

$$\min U(f_0' , f_0'') \approx \min U(H'^t, H''^t) \approx \min U(f_1' , f_1'')$$

as to be asserted.

Consider now the union $K = K' \underset{L}{\cup} K''$ of two complexes K', K'' along a common subcomplex L with inclusions $i' : L \subset K'$, $i'' : L \subset K''$ as in the beginning of this section. Let

$$\mathcal{A}' = A^*(K') , \quad \mathcal{A}'' = A^*(K'') , \quad \mathcal{B} = A^*(L) ,$$

$$\mathcal{M}' = I^*(K') , \quad \mathcal{M}'' = I^*(K'') , \quad \mathcal{N} = I^*(L)$$

and $g : \mathcal{N} \to \mathcal{B}$ be some associated minimal morphism. By theorem in § II.5 we can then define, in starting from i'^A, i''^A and g, some associated minimal morphisms $g' : \mathcal{M}' \to \mathcal{A}'$, $g'' : \mathcal{M}'' \to \mathcal{A}''$ and also i'^I, i''^I to get a strictly commutative diagram below

$$\begin{array}{ccccc}
\mathcal{A}' & \xrightarrow{i'^A} & \mathcal{B} & \xleftarrow{i''^A} & \mathcal{A}'' \\
{\scriptstyle g'}\uparrow & & \uparrow{\scriptstyle g} & & \uparrow{\scriptstyle g''} \\
\mathcal{M}' & \xrightarrow{i'^I} & \mathcal{N} & \xleftarrow{i''^I} & \mathcal{M}''
\end{array}$$

Now, both i'^A, i''^A are onto, but i'^I, i''^I are generally not so that Prop. 2 cannot be applied. In order to apply Prop. 2 we now use the following device.

<u>Definition</u>: For any free DGA $\mathcal{F}$ with free generators u_i let $\mathcal{F}^{tr}$ be the free DGA with free generators u_i^0, u_i^+, and degrees and differentials below:

$$\deg u_i^0 = \deg u_i , \qquad \deg u_i^+ = \deg u_i + 1 ,$$

$$du_i^0 = u_i^+ , \qquad du_i^+ = 0 .$$

Define also a DGA-morphism

$$tr : \mathcal{F}^{tr} \to \mathcal{F}$$

by $tr(u_i^0) = u_i$, $tr(u_i^+) = du_i$. The DGA $\mathcal{F}^{tr}$ will be called the <u>trivialization</u> of $\mathcal{F}$ and tr the corresponding <u>trivializing-morphism</u>.

<u>Definition</u>: For any DGA's $\mathcal{F}, \mathcal{G}$ with $\mathcal{F}$ free and a DGA-morphism

$$g : \mathcal{G} \to \mathcal{F}$$

we set $\mathcal{G} \otimes \mathcal{F}^{tr} = \mathcal{G}_{\mathcal{F}}$. The DGA-morphism

$$g_{\mathcal{F}} = g \otimes tr : \mathcal{G}_{\mathcal{F}} \to \mathcal{F}$$

will then be called the <u>trivial extension</u> of g , which is an epimorphism.

<u>Remark</u>: I owe this trick of turning a DGA morphism to an epimorphism by introducing trivial extensions, to Prof. J. Morgan of Columbia University in New York.

<u>Theorem 1</u>: The I^*-measure is calculable with respect to the union construction of two complexes along a common subcomplex. More precisely, let $K = K' \underset{L}{\cup} K''$ be the union of K' , K" along a common subcomplex L with inclusions

$$i' : L \subset K' , \quad i'' : L \subset K'' .$$

Set $\mathcal{M}' = I^*(K')$, $\mathcal{M}'' = I^*(K'')$, $\mathcal{N} = I^*(L)$ and replace the diagram

$$(D_I) \qquad \mathcal{M}' \xrightarrow{i'^I} \mathcal{N} \xleftarrow{i''^I} \mathcal{M}''$$

by the diagram

$$\mathcal{M}'_{\mathcal{N}} \xrightarrow{i'^I_{\mathcal{N}}} \mathcal{N} \xleftarrow{i''^I_{\mathcal{N}}} \mathcal{M}''_{\mathcal{N}}$$

in forming trivial extensions. Then

$$I^*(K) \simeq \min \cup (i'^I_{\mathcal{N}}, i''^I_{\mathcal{N}}) .$$

<u>Proof</u>: Let $\mathcal{A}' = A^*(K')$, etc. and consider the commutative diagram

$$\begin{array}{ccccc} \mathcal{A}' & \xrightarrow{i'^A} & \mathcal{B} & \xleftarrow{i''^A} & \mathcal{A}'' \\ g' \uparrow & & \uparrow g & & \uparrow g'' \\ \mathcal{M}' & \xrightarrow{i'^I} & \mathcal{N} & \xleftarrow{i''^I} & \mathcal{M}'' \end{array}$$

as before. Since i'^A , i''^A both are onto, we can clearly extend g' , g" to DGA-morphisms

$$\tilde{g}' : \mathcal{M}'_{\mathcal{N}} \to \mathcal{A}' ,$$

$$\tilde{g}'' : \mathcal{M}''_{\mathcal{N}} \to \mathcal{A}''$$

so that the following diagram is commutative

$$\begin{array}{ccccc} \mathcal{A}' & \xrightarrow{i'^{A}} & \mathcal{B} & \xleftarrow{i''^{A}} & \mathcal{A}'' \\ \uparrow{\scriptstyle \tilde{g}'} & & \uparrow{\scriptstyle g} & & \uparrow{\scriptstyle \tilde{g}''} \\ \mathcal{M}'_{\mathcal{N}} & \xrightarrow{i'^{I}_{\mathcal{N}}} & \mathcal{N} & \xleftarrow{i''^{I}_{\mathcal{N}}} & \mathcal{M}''_{\mathcal{N}} \end{array}$$

From Prop. 1 and 2 we have then

$$I^*(K) \approx \min U (i'^{A}, i''^{A})$$

$$\approx \min U (i'^{I}, i''^{I}) ,$$

as to be proved.

<u>Remark</u>: The DGA-morphisms i'^{I}, i''^{I} are unique only up to homotopy. However, they will lead to the same $I^*(K)$ in view of Prop. 3 .

<u>Theorem 2</u>: Let $K = K' \underset{L}{\cup} K''$ be the union of K', K'' along L as in Theorem 1 . Then the minimal morphisms

$$I^*(K) \to I^*(K') \quad \text{resp.} \quad I^*(K'')$$

induced by the inclusions $i' : K' \subset K$, $i'' : K'' \subset K$ are completely determined by the diagram

$$(D_I) \qquad I^*(K') \xrightarrow{i'^{I}} I^*(L) \xleftarrow{i''^{I}} I^*(K'') .$$

<u>Proof</u>: Using the notations as before $U(i'^{A}, i''^{A})$ is a sub-DGA of $A^*(K') \oplus A^*(K'')$.

Then the projection p' of $A^*(K') \oplus A^*(K'')$ onto $A^*(K')$, when restricted on $U(i'^{A}, i''^{A})$ will give rise to a DGA-morphism, still denoted by p' , viz.

$$p' : U(i'^{A}, i''^{A}) \to A^*(K') .$$

In a similar manner, there will be a DGA-morphism induced by inclusion followed by projection, say $(\mathcal{N} = I^*(L))$

$$q' : U(i'^{I}_{\mathcal{N}}, i''^{I}_{\mathcal{N}}) \to \mathcal{M}'_{\mathcal{N}}.$$

We will clearly have a commutative diagram

$$\begin{array}{ccc} A^*(K) \approx U(i'^A, i''^A) & \xrightarrow{p'} & A^*(K') \\ \uparrow{\scriptstyle \tilde{g}} & & \uparrow{\scriptstyle \tilde{g}'} \\ U(i'^I_{\mathcal{N}}, i''^I_{\mathcal{N}}) & \xrightarrow{q'} & \mathcal{M}'_{\mathcal{N}} \end{array}$$

in which $\tilde{g}'$ and $\tilde{g} = (\tilde{g}', \tilde{g}'')$ are as before. Since

$$\min \mathcal{M}'_{\mathcal{N}} \approx \min \mathcal{M}' \approx I^*(K')$$

and

$$\min U(i'^I_{\mathcal{N}}, i''^I_{\mathcal{N}}) \approx I^*(K)$$

by Theorem 1, we clearly get the present Theorem.

Corollary: For the union $K = K' \underset{L}{\cup} K''$ of K', K'' along L, as before, the cohomology ring H^*-measure $H^*_{\underline{k}}(K) \approx H^*(K)$ of K on $\underline{k}$ is completely determined by diagram (D_I), i.e. by the $\underline{I}^*$-measures of K', K'', and L as well as the associated minimal DGA-morphisms induced from the inclusions of L in K', K'':

$$H^*(K) \approx H(U(i'^I_{\mathcal{N}}, i''^I_{\mathcal{N}})) .$$

Remark: As shown by examples in § I.3, the $H^*(K)$ cannot be determined by the corresponding knowledge in H^* alone, viz. from the diagram (D_H) below alone:

$$(D_H) \qquad H^*(K') \xrightarrow{i'^H} H^*(L) \xleftarrow{i''^H} H^*(K'') .$$

VI.3 SOME PARTICULAR CASES - CONE-CONSTRUCTION AND SUSPENSION

Let us consider the particular case of a union

$$K = K' \underset{L}{\cup} K''$$

of two complexes K', K'' along a common sub-complex L for which K'' is the cone over L, by adjoining a new vertex, say v. In this case, $I^*(K'')$ is trivial and the determination of $I^*(K)$ from the diagram

$$I^*(K') \xrightarrow{i'^I} I^*(L) \xleftarrow{i''^I} I^*(K'') \approx \underline{k}$$

becomes much simpler than that one given in § VI.2. To explain it, let us first introduce some general notions.

<u>Definition</u>: For any DGA's $\mathcal{A},\mathcal{B}$ and a DGA-morphism

$$f : \mathcal{A} \to \mathcal{B}$$

let $\mathcal{C}_0$ = the basic field $\underline{k}$, and

$$\mathcal{C}_p = \mathrm{Ker}\,[f : \mathcal{A}_p \to \mathcal{B}_p] , \qquad p > 0 .$$

With algebraic operations, degree and differential inherited from those of $\mathcal{A}$, the direct sum $\sum_{p\geq 0} \mathcal{C}_p$ naturally will become a sub-DGA of $\mathcal{A}$ which will be called the <u>kernel</u> of f and will be denoted by $\mathrm{Ker}\, f$.

Consider now a pair of a complex K and a subcomplex L with inclusion $i : L \subset K$. As in § I.3, let us form the cone-construction by erecting a cone C_L over L by adjoining some new vertex and forming the union $K \underset{L}{\cup} C_L$ which is denoted by $\Delta_L(K)$ or simply Δ in what follows. Let $j : L \subset C_L$, $\tilde{j} : C_L \subset \Delta$, $\tilde{i} : K \subset \Delta$ be all inclusions. We have then the following

<u>Proposition</u>: Let $\Delta = \Delta_L(K)$ be the union $K \underset{L}{\cup} C_L$ with C_L the cone erected over a subcomplex L of K. Then

$$I^*(\Delta) \simeq \min \mathrm{Ker}\, i^A .$$

Moreover, the minimal morphisms

$$I^*(\Delta) \to I^*(K)$$

associated to the inclusion

$$\tilde{i} : K \subset \Delta$$

are induced from the natural inclusion

$$\mathrm{Ker}\, i^A \subset A^*(K) .$$

<u>Proof</u>: Let us consider the following diagram of DGA's and DGA-morphisms:

$$\begin{array}{ccccccccc}
 & & & & A^*(\Delta) & \xrightarrow{\tilde{j}^A} & A^*(C_L) & \longrightarrow & 0 \\
 & & & \overset{\tau}{\nearrow} & \downarrow{\scriptstyle \tilde{i}^A} & & \downarrow{\scriptstyle j^A} & & \\
0 & \longrightarrow & \operatorname{Ker} i^A & \longrightarrow & A^*(K) & \xrightarrow[i^A]{} & A^*(L) & \longrightarrow & 0
\end{array}$$

In the diagram, by the Extension Theorem of § VI.1 , the two horizontal lines both are exact sequences, except for the case of degree 0 for the lower horizontal line. The morphism τ is to be defined as follows.

Consider any $x \in \operatorname{Ker} i^A$. If $\deg x = 0$ or, that x is an element of the basic field $\underline{k}$, then $\tau x \in A^0(\Delta)$ will be that differential form which will take on the constant-valued 0-form x on each simplex in Δ. For $\deg x > 0$, the restriction $i^A x$ will be a form 0 on each simplex in L. Hence, we naturally may extend x to a form in $A^*(\Delta)$ such that it is a form 0 on each simplex in C_L. It is clear that τ induces a DGA-morphism which makes the above diagram a commutative one.

Now, let us prove that τ is an H-isomorphism. In fact, in degree 0 there is nothing to prove. Consider now a form $\tilde{x} \in A^*(\Delta)$ with $\deg \tilde{x} > 0$ and $d\tilde{x} = 0$. Since C_L is homologically trivial, there will be some $y \in A^*(C_L)$ such that $\tilde{j}^A \tilde{x} = dy$ in C_L. Since $\tilde{j}^A$ is onto, there will be some $\tilde{y} \in A^*(\Delta)$ such that $\tilde{j}^A \tilde{y} = y$. Set $\tilde{z} = \tilde{x} - d\tilde{y}$, then $\tilde{j}^A \tilde{z} = 0$. Hence, $i^A \tilde{i}^A \tilde{z} = j^A \tilde{j}^A \tilde{z} = 0$ or $\tilde{i}^A \tilde{z} \in \operatorname{Ker} i^A$.
Now, $\tau \tilde{i}^A \tilde{z} = \tilde{z} = \tilde{x} - d\tilde{y} \sim \tilde{x}$ and $d\tilde{i}^A \tilde{z} = 0$. Consequently τ_H is an epimorphism.

Consider now any $z \in \operatorname{Ker} i^A$ with $\deg z > 0$, $dz = 0$ and $\tau z = d\tilde{a}$ for some $\tilde{a} \in A^*(\Delta)$. Since $d\tilde{j}^A \tilde{a} = \tilde{j}^A d\tilde{a} = \tilde{j}^A \tau z = 0$, C_L is homologically trivial, and $\tilde{j}^A$ is onto, there exists $\tilde{b} \in A^*(\Delta)$ with $\tilde{j}^A \tilde{a} = d\tilde{j}^A \tilde{b}$. Then, $\tilde{i}^A(\tilde{a} - d\tilde{b}) \in \operatorname{Ker} i^A$ and $z = d\tilde{i}^A(\tilde{a} - d\tilde{b}) \sim 0$ in $\operatorname{Ker} i^A$. Hence, τ_H is a monomorphism.

It follows that τ is an H-isomorphism so that

$$I^*(\Delta) \approx \min A^*(\Delta) \approx \min \operatorname{Ker} i^A .$$

Furthermore, the minimal morphisms $I^*(\Delta) \to I^*(K)$ will be induced from the inclusion $\operatorname{Ker} i^A \subset A^*(K)$ since they are induced from $\tilde{i}^A : A^*(\Delta) \to A^*(K)$.

q.e.d.

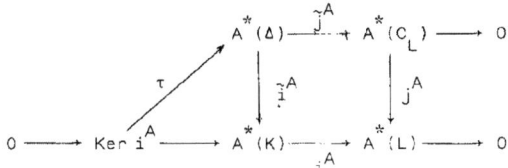

Similar to the proof of Theorem 1 in the last section, we get from the above Proposition the following

Theorem 1: The I^*-measure is calculable with respect to the cone-construction over a subcomplex. More precisely, with the same notations as in the Proposition, the I^*-measure of $\Delta = \Delta_L(K)$ is completely determined from

$$i^I : I^*(K) \to I^*(L) \approx \mathcal{N}$$

as

$$I^*(\Delta) \approx \min \operatorname{Ker} i^I_{\mathcal{N}} ,$$

where

$$i^I_{\mathcal{N}} = (i^I)^{tr} : I^*(K)_{\mathcal{N}} = I^*(K)_{I^*(L)} \to I^*(L)$$

is the DGA-epimorphism, as defined in § VI.2 . Moreover, the minimal morphisms

$$I^*(\Delta) \to I^*(K)$$

will be induced from the inclusion $\operatorname{Ker} i^I_{\mathcal{N}} \subset I^*(K)_{I^*(L)}$. Since $\min I^*(K)_{I^*(L)} \approx I^*(K)$.

Notation: Henceforth, the minimal morphism $I^*(\Delta) \to I^*(K)$ will be denoted by $\gamma(i)$ which is deduced from the inclusion $i : L \subset K$.

From the proof we see that the following theorem is true:

Theorem 2: The determination of $I^*(\Delta_L(K))$ from $I^*(K)$ is functorial in the sense that for complexes $L \subset K \subset K'$ with inclusions $i : L \subset K$, $i' : L \subset K'$ we have a $\approx$-commutative or even commutative diagram of DGA-morphisms

$$\begin{array}{ccccc} I^*(\Delta_L(K)) & \xrightarrow{\gamma(i)} & I^*(K) & \xrightarrow{i^I} & I^*(L) \\ {\scriptstyle \tilde{j}^I}\uparrow & & \uparrow{\scriptstyle j^I} & & \| \\ I^*(\Delta_L(K')) & \xrightarrow[\gamma(i')]{} & I^*(K') & \xrightarrow[i'^I]{} & I^*(L) \end{array}$$

in which $j : K \subset K'$, $\tilde{j} : \Delta_L(K) \subset \Delta_L(K')$ are also inclusions

As a corollary of Theorem we further have the following

<u>Theorem 3</u>: The I^*-measure is calculable with respect to the suspension-construction for connected complexes. More precisely, let $\tilde{K}$ be the suspension of a connected complex K. Then, $I^*(\tilde{K})$ is completely determined by $I^*(K)$ as

$$I^*(\tilde{K}) \approx \min \operatorname{Ker} tr \ ,$$

where

$$tr : I^*(K)^{tr} \to I^*(K)$$

is the trivializing-morphism of $I^*(K)$.

Below, we give some example to illustrate the applications of the preceding Theorems.

<u>Example</u>: I^*-measure of a sphere.

<u>Proposition</u>: For a sphere S^n of dimension $n \geq 1$ the I^*-measure is given by:

For n odd,

$$I^*(S^n) \approx \operatorname{Extr}(x)$$

with

$$\deg x = n \quad \text{odd}, \qquad dx = 0 \ .$$

For n even,

$$I^*(S^n) \approx \operatorname{Polym}(y) \otimes \operatorname{Extr}(x)$$

with

$$\deg y = n \text{ even}, \quad \deg x = 2n-1,$$
$$dy = 0, \qquad dx = y^2.$$

Proof: For $n = 1$ it is trivial. We prove the general case by induction on n. In fact, let S^{n-1} be the equator-sphere of S^n, then S^n is the suspension of S^{n-1}, so that Theorem 3 may be applied.

To facilitate the calculations let us form a DGA J_{n-1} of a single generator s with

$$\deg s = n-1, \quad ds = 0, \quad \text{and} \quad s^2 = 0.$$

It is clear that the minimal model of J_{n-1} is $I^*(S^{n-1})$, and we may apply Theorem 3 to J_{n-1} instead of $I^*(S^{n-1})$ to get $I^*(S^n)$. This will greatly simplify the calculations in the case n = odd.

Construct thus the trivialization of J_{n-1} and the corresponding trivializing morphism, viz.

$$tr : \mathrm{Free}(s^0, s^+) \to J_{n-1}$$

with

$$\deg s^0 = n-1, \quad \deg s^+ = n,$$
$$ds^0 = s^+, \quad ds^+ = 0,$$
$$tr(s^0) = s, \quad tr(s^+) = 0.$$

The positive degree part of the DGA Ker tr considered as a module, has then a basis below.

For n odd:

$$s^+, \; s^+s^0, \; s^+(s^0)^2, \; \ldots, \; (s^0)^2, \; (s^0)^3, \; \ldots .$$

For n even:

$$s^+, \; (s^+)^2, \; (s^+)^3, \; \ldots, \; s^0s^+, \; s^0(s^+)^2, \; \ldots .$$

Now, in case n odd, we have

$$d(s^0)^k = k\,s^+(s^0)^{k-1}, \qquad k \geq 2$$

and in case n even, we have

$$d(s^0(s^+)^k) = (s^+)^{k+1}, \quad k \geq 1 .$$

Hence, in any case the minimal model of Ker tr is easily seen to be the same as that of the DGA consisting of a single generator t with

$$\deg t = n, \quad dt = 0, \quad t^2 = 0 .$$

The corresponding minimal model is clearly $I^*(S^n)$, as described above.

VI.4 EFFECTIVE COMPUTATION AND AXIOMATIC SYSTEM OF I^*-MEASURE

The 1-dimensional skeleton K_0 of a connected finite complex K has the same homotopy type as the union of a finite set of 1-spheres, joined at a single point. By either Chap. III or the preceding sections, the I^*-measure of K_0 can be computed effectively. As any finite complex can be obtained from its 1-dimensional skeleton by successively adding higher-dimensional simplexes of which each step is homotopically equivalent to a cone construction, Theorem 1 of § VI.2 furnishes us with a means to compute the I^*-measure of any connected finite complex effectively.

In more details, let us represent any connected finite simplicial complex K in the form

$$K = K_m \supset \ldots \supset K_1 \supset K_0 .$$

In this sequence, K_0 is the 1-dimensional skeleton of K, and K_r is the union of K_{r-1} with an additional simplex Δ_r, the boundary of which is

$$\dot{\Delta}_r = L_{r-1} \subset K_{r-1}$$

so that

$$K_r = K_{r-1} \underset{L_{r-1}}{\cup} \Delta_r = \Delta_{L_{r-1}}(K_{r-1})$$

in the notation of § VI.3 .

Let

$$i_{r-1} : L_{r-1} \subset K_{r-1}$$

be the inclusion and

$$i^I_{r-1} : I^*(K_{r-1}) \to I^*(L_{r-1})$$

be any of the associated morphisms of I^*-measures. Now, we have the following

Lemma: If we know how to compute $I^*(L'_{r-1})$, $I^*(L''_{r-1})$ and the associated DGA-morphisms

$$k^I_{r-1} : I^*(L'_{r-1}) \to I^*(L''_{r-1})$$

for any connected subcomplexes $L''_{r-1} \subset L'_{r-1}$ of K_{r-1}, with k_{r-1} as the inclusion, then we also know how to compute $I^*(L'_r)$, $I^*(L''_r)$ and the associated DGA-morphisms

$$k^I_r : I^*(L'_r) \to I^*(L''_r)$$

for any connected subcomplex $L''_r \subset L'_r$ of K_r with inclusion k_r.

Proof: We consider three cases separately.

Case 1: $L''_r \subset L'_r \subset K_{r-1}$. In this case, $I^*(L'_r)$, $I^*(L''_r)$ and $k^I_r : I^*(L'_r) \to I^*(L''_r)$ are already known by hypothesis.

Case 2: $L''_r \subset K_{r-1}$ but $L'_r \not\subset K_{r-1}$. In that case, L'_r contains Δ_r as a sub-simplex and may be considered as the cone construction over L_{r-1} of some sub-complex $\bar{L}'_r$ of K_{r-1} ($L'_r = \Delta_{L_{r-1}}(\bar{L}'_r)$), while L''_r is a subcomplex of $\bar{L}'_r$. It follows from Theorem 1 of § VI.3 that $I^*(L'_r)$ is determined by the DGA-morphism

$$j^I_r : I^*(\bar{L}'_r) \to I^*(L_{r-1}),$$

already known by hypothesis, with $j_r : L_{r-1} \subset \bar{L}'_r$ the inclusion. The associated DGA-morphisms

$$I^*(L'_r) \to I^*(L''_r)$$

are then the composition

$$I^*(L'_r) \xrightarrow{\gamma(j_r)} I^*(\bar{L}'_r) \longrightarrow I^*(L''_r)$$

of which the morphism on the right is again known by hypothesis.

Case 3: $L''_r \not\subset K_{r-1}$. In this case, L'_r, L''_r both contain Δ_r as a sub-simplex and can be considered as cone constructions over L_{r-1}:

$$L''_r = \Delta_{L_{r-1}}(\bar{L}''_r), \quad L'_r = \Delta_{L_{r-1}}(\bar{L}'_r).$$

Now, $\bar{L}''_r \subset \bar{L}'_r \subset K_{r-1}$, so that $I^*(\bar{L}'_r)$, $I^*(\bar{L}''_r)$ and $\bar{k}^I_{r-1} : I^*(\bar{L}'_r) \to I^*(\bar{L}'_r)$ with inclusion $\bar{k}_{r-1} : \bar{L}''_r \subset \bar{L}'_r$ are known by hypothesis; thus

$$I^*(L'_r) , \quad I^*(L''_r) \quad \text{and} \quad k^I_r : I^*(L'_r) \to I^*(L''_r)$$

can be determined by Theorem 2 of § VI.3 .

In view of § III.5 , the theorems in the preceding sections and the Lemma above we have the following

Theorem 1: There is an algorithmic procedure permitting us to compute effectively the I^*-measure of any connected finite complex up to any prescribed degree.

The same considerations also furnish us immediately with an axiomatic system for the I^*-measure over the category of connected finite complexes (or polytopes) as follows.

To each complex K in the category a minimal DGA $I^*(K)$ is associated, and to each pair of complex K and subcomplex K' with inclusion $k : K' \subset K$ there is associated a set of DGA-morphisms determined up to homotopy:

$$k^I : I^*(K) \to I^*(K') .$$

These DGA algebras and morphisms are characterized by the following axiomatic system (Wu[6,9]) :

Axiom 1: I^* is a homotopy measure, in other words, for K, K' having the same homotopy type in the category, we have $I^*(K) \approx I^*(K')$.

Axiom 2: If $K'' \subset K' \subset K$ are complexes, all in the above category and

$$k^I : I^*(K) \to I^*(K') , \qquad k'^I : I^*(K') \to I^*(K'')$$

are associated DGA-morphisms for the pairs $K' \subset K$ and $K'' \subset K'$, then $k'^I k^I : I^*(K) \to I^*(K'')$ are also associated DGA-morphisms for the pair $K'' \subset K$.

Axiom 3: If $\Delta = \Delta_L(K)$ is the cone construction over a subcomplex L of K, both being in the above category, then $I^*(\Delta)$ is given by

$$I^*(\Delta) \approx \min \operatorname{Ker} \tilde{j}^I ,$$

where

$$j^I : I^*(K) \to I^*(L)$$

is any of the associated DGA-morphisms of the pair $L \subset K$ with inclusion j, and $\tilde{j}^I = j^I \otimes tr : I^*(K) \otimes I^*(L)^{tr} \to I^*(L)$ is determined by j^I.

Axiom 4: For the cone construction $\Delta = \Delta_L(K)$, as in Axiom 3, the associated DGA-morphisms of the pair $K \subset \Delta$ with inclusion j are given by the natural DGA-morphisms

$$\gamma(j) : I^*(\Delta) \to I^*(K) .$$

Axiom 5: The I^*-measure $I^*(\Delta)$ and the morphism $\gamma(j)$ in Axioms 3 and 4 are functorial in character. In other words, if we have pairs $L \subset K$ and $L' \subset K'$ in the category with L', K' as subcomplexes of L and K respectively, then we have the following homotopically commutative diagram with evident DGA-morphisms:

$$\begin{array}{ccc} I^*(\Delta) & \longrightarrow & I^*(K) \\ \downarrow & & \downarrow \\ I^*(\Delta') & \longrightarrow & I^*(K') \end{array}$$

Axiom 6: For any two complexes K', K'' in the category we have

$$I^*(K' \vee K'') \simeq \min(I^*(K') + I^*(K''))$$

and the associated DGA-morphisms

$$I^*(K' \vee K'') \to I^*(K')$$

are determined by the composition

$$I^*(K' \vee K'') \to I^*(K') \oplus I^*(K'') \to I^*(K') .$$

Axiom 7: For a 1-sphere S' the I^*-measure is given by

$$I^*(S') \simeq \text{Free}(x) \quad \text{with}$$

$$\deg x = 1 , \quad dx = 0 .$$

We have restricted our considerations to finite complexes which, however, can easily be extended to infinite ones, in the following manner.

Let K be a connected complex which is the union of an increasing sequence of connected finite complexes:

$$K_1 \underset{i_1}{\subset} K_2 \underset{i_2}{\subset} \dots \subset K_m \underset{i_m}{\subset} \dots ,$$

the i'_ms all being inclusions. We have then a sequence of DGA's and DGA-morphisms for the A^*-measures:

$$(D_A) \qquad A^*(K_1) \xleftarrow{i_1^A} A^*(K_2) \xleftarrow{i_2^A} \cdots \longleftarrow A^*(K_m) \xleftarrow{i_m^A} \cdots$$

By § VI.1 , all the DGA-morphisms i_m^A are onto.
Hence, by § II.5 we can complete the sequence (D_A) to a strictly commutative diagram (D_I^A) of DGA-morphisms shown below:

$$(D_I^A) \qquad \begin{array}{ccccccccc} A^*(K_1) & \xleftarrow{i_1^A} & A^*(K_2) & \xleftarrow{i_2^A} & \cdots & \longleftarrow & A^*(K_m) & \xleftarrow{i_m^A} & \cdots \\ \uparrow{\scriptstyle \rho_1} & & \uparrow{\scriptstyle \rho_2} & & & & \uparrow{\scriptstyle \rho_m} & & \\ I^*(K_1) & \xleftarrow[i_1^I]{} & I^*(K_2) & \xleftarrow[i_2^I]{} & \cdots & \longleftarrow & I^*(K_m) & \xleftarrow[i_m^I]{} & \cdots \end{array}$$

In the diagram (D_I^A) all ρ_m are respective minimal morphisms. The I^*-measure of the infinite complex K is then just the minimal model of the inverse limit as a DGA of the sequence

$$(D_I) \qquad I^*(K_1) \xleftarrow[i_1^I]{} I^*(K_2) \xleftarrow[i_2^I]{} \cdots \longleftarrow I^*(K_m) \xleftarrow[i_m^I]{} \cdots$$

This is easily proved, since we are always working in the case of fields of characteristic 0 . Cf. Eilenberg-Steenrod, Foundations of Algebraic Topology, Vol. 1 Chap. VIII, § 6, cf. also Kahn [1] .

VI.5 SOME APPLICATIONS TO FIBRATIONS - FIBER SPACE THEOREM

We consider only locally trivial fibrations.

Consider a fibration

$$(\mathcal{F}) \qquad F \subset E \xrightarrow{g} B$$

in which F is the fiber, E the fiber space, and B the base space. We shall suppose that the fibration is <u>simplicial</u> in the following sense.

<u>Definition</u>: The fibration $(\mathcal{F})$ is called <u>simplicial</u> if all the spaces F, E, B are simplicial complexes, if the inclusion $i : F \subset B$ and the projection $g : E \to B$ both are simplicial maps, and if $\pi_1(B)$ acts trivially on the homology of fibers.

Remark: It seems that the fibration under consideration is quite restrictive. However, it already contains the important case of differential fibration in which all the spaces are differentail manifolds, and i and g both are differential, owing to known theorems of J.H.C. Whitehead [1] on triangulations of differential manifolds. A second important example of simplicial fibration is furnished by the $K(\pi, n-1)$ - bundle induced by the universal fibration

$$(\mathcal{K}) \qquad K(\pi, n-1) \underset{i}{\subset} P(\pi,n) \xrightarrow[g]{} K(\pi,n)$$

in which π is a countable abelian group and $P(\pi,n)$ is contractible. In fact, according to Milgram [1] and Steenrod [2], we can realize the above fibration with all spaces involved by CW-complexes and all maps, the inclusion i and projection g, as skeletal ones. As already pointed out in § I.4, $K(\pi, n-1)$ and $K(\pi,n)$ both possess simplicial subdivisions. We can then easily subdivide $P(\pi,n)$ simplicially, such that $K(\pi, n-1)$ will become a simplicial subcomplex of $P(\pi,n)$ and g will also be a simplicial map. The fibration $(\mathcal{K})$ becomes then a simplicial one, as desired. Let B be any simplicial complex and $f : B \to K(\pi,n)$ be a continuous map. By further subdivision of B and simplicial approximation of f by some simplicial map s, the induced fibration $f^*\mathcal{K}$ over B will then be homotopically equivalent to the simplicial fibration $s^*\mathcal{K}$ over B.

The simplicial fibrations are thus quite usual, as well as natural. Moreover, theorems proved for such simplicial fibrations can easily be extended to more general cases of various kinds of fibrations by either following the same pattern of proofs exhibited for the simplicial ones or by simple generalizations depending mainly on different kinds of homology theory which have been used. So we shall restrict our considerations to simplicial fibrations in this and the next section as well as in the following chapter.

One of the most important problems concerning the study of a fibration is the determination of a certain measure of the fiber space E in terms of those of the fiber F, and the base B. In the case of the cohomology-ring H^*-measure, a powerful method frequently used is furnished by various kinds of spectral sequences connected with the fibration. When the spectral sequences collapse, which will only occur under quite restrictive conditions, then we may achieve the end by arriving at a complete determination of the H^*-measure in question. However, in general this is not the case, and it is even impossible in view of the calculability. An alternative method of dealing with the problem is suggested by E.H. Brown [1] by proving that

(1) $$H_{\oplus}(E) \approx H(S(B) \underset{\tau}{\otimes} S(F))$$

in which S(X) means the group of singular chains of a space X and $\underset{\tau}{\otimes}$ means certain somewhat <u>twisted</u> boundary operators have been introduced in the tensor product. Though the expression is explicit and concise, it can hardly be applied, since the measure S is not an adequate one in view of the second requirement discussed in § I.3 . However, if we replace the S-measure by the I^*-measure then we shall get for the complete determination of the I^*-measure, and hence also the H^*-measure, of E in terms of the I^*-measures (but not at all the H^*-measures) of B and F in an explicit form, similar to that of E.H. Brown. In fact, we have the following

<u>Theorem (Fiber Space Theorem</u>: For a simplicial fibration $(\mathcal{F})$ the I^*-measure of the fiber space E is given by

(2) $$I^*(E) \approx \min (I^*(B) \underset{\tau}{\otimes} I^*(F)) ,$$

in which $I^*(B) \underset{\tau}{\otimes} I^*(F)$ is the tensor product with a certain <u>twisted</u> differential

(3) $$d = d_{\otimes} + d_{\tau} .$$

Here $d_{\otimes} = d_B \otimes 1 \pm 1 \otimes d_F$ is the usual differential as a tensor product determined by the differentials d_B in $I^*(B)$ and d_F in $I^*(F)$, while d_τ is an additional <u>twisted</u> part.

<u>Remark</u>: The Fiber-Space Theorem gives the determination of the I^*-measure of the fiber space in terms of those of the base B and the fiber F in an explicit and also concise form. The only defect is, that for the twisted differential d_τ only the existence has been shown, but no constructive means has been indicated. In certain important cases the twisted differential can in fact be given constructively, which, however, will be postponed to the next chapter, besides the case considered in the next section.

<u>Proof</u>: In view of the remarks made at the end of the last section we shall restrict ourselves to the case of B being a finite connected complex, considered one obtained by adjoining successive simplexes. Thus, we shall prove the theorem by induction, which can easily seen to be true for B connected and of dimension 1 . Now, let B be the union $B_1 \cup \Delta$ where $\Delta = B_2$ is a simplex of dimension n adjoined to B_1 with its boundary $\dot{\Delta} = B_0$ an (n-1)-dimensional combinatorial sphere in B_1 , and suppose that the theorem has already been proved for the fi-

bration over B_1. We have to show that the theorem remains true for the fibration over B.

Denote for this purpose the parts of the fibration $(\bar{F})$ over B_1, $\dot{\Delta}$, Δ by $(\bar{F}_1)$, $(\bar{F}_0)$, $(\bar{F}_2)$ with fiber spaces E_1, E_0, E_2 and projections g_1, g_0, g_2 respectively. Then, we have to consider the ≃-commutative diagram below:

$$(D)\qquad \begin{array}{ccccc} I^*(E_1) & \xrightarrow{\tilde{f}_1^I} & I^*(E_0) & \xleftarrow{\tilde{f}_2^I} & I^*(E_2) \\ \uparrow g_1^I & & \uparrow g_0^I & & \uparrow g_2^I \\ I^*(B_1) & \xrightarrow{f_1^I} & I^*(B_0) & \xleftarrow{f_2^I} & I^*(B_2) \end{array}$$

In the diagram f_i, $\tilde{f}_i$ are all inclusions, and $B_0 = \dot{\Delta}$, $B_2 = \Delta$. By induction we have

$$(4)_1 \qquad I^*(E_1) \simeq I^*(B_1) \otimes_{\tau_1} I^*(F)$$

with a twisted differential $d_{\otimes} + d_{\tau_1}$, while the differentials in

$$(4)_0 \qquad I^*(E_0) \simeq I^*(B_0) \otimes I^*(F) \quad \text{and}$$

$$(4)_2 \qquad I^*(E_2) \simeq I^*(B_2) \otimes I^*(F) \simeq I^*(F)$$

are ordinary ones, since the fibrations $(\bar{F}_0)$ and $(\bar{F}_2)$ both are trivial.

By the theorems in § VI.2 the I^*-measure of B is to be determined by the lower horizontal line of diagram (D) and that of $E = E_1 \underset{E_0}{\cup} E_2$ by the upper line. To facilitate the calculations we shall introduce some auxiliary DGA's, as follows.

Let S be the DGA with a single generator s such that

$$(5) \qquad s^2 = 0, \quad ds = 0, \quad \deg s = n-1.$$

Let J_i $(i = 1,2)$ be the DGA with two generators s_i and t_i such that

$$(6) \qquad \begin{cases} s_i^2 = 0, & s_i t_i = 0, \quad t_i^2 = 0, \\ ds_i = t_i, & dt_i = 0, \\ \deg s_i = n-1, & \deg t_i = n. \end{cases}$$

Then, J_i are homologically trivial and

$$(7)\qquad tr_i : \begin{cases} s_i \to s , \\ t_i \to 0 \end{cases}$$

are DGA-morphisms of J_i onto $\mathcal{S}$. The minimal model of $\mathcal{S}$ is $I^*(B_0) = I^*(\dot{\Delta})$ and the associated minimal morphism $I^*(B_0) \to \mathcal{S}$ will be denoted by ρ_0. For $i = 1,2$, let us write for simplicity $f_i = \rho_0 f_i^I$, $\tilde{f}_i = (\rho_0 \otimes 1)\tilde{f}_i^I$ and

$$(8)\qquad \begin{cases} J_i \otimes I^*(B_i) = \mathcal{A}_i , \quad \mathcal{S} = \mathcal{A}_0 , \\ J_i \otimes I^*(E_i) = \tilde{\mathcal{A}}_i , \quad \mathcal{S} \otimes I^*(F) = \tilde{\mathcal{A}}_0 , \\ tr_i \otimes f_i = f_i^{tr} : \mathcal{A}_i \to \mathcal{A}_0 , \\ tr_i \otimes \tilde{f}_i = \tilde{f}_i^{tr} : \tilde{\mathcal{A}}_i \to \tilde{\mathcal{A}}_0 . \end{cases}$$

By theorems in the preceding sections we have then

$$(9)\qquad I^*(B) \approx \min U(f_1^{tr}, f_2^{tr}) ,$$

$$(10)\qquad I^*(E) \approx \min U(f_1^{tr}, f_2^{tr}) .$$

in which $U(f_1^{tr}, f_2^{tr})$ and $U(\tilde{f}_1^{tr}, \tilde{f}_2^{tr})$ are to be determined respectively by the two diagrams below.

$$(D_0)\qquad \begin{array}{ccccc} \mathcal{A}_1 & \xrightarrow{f_1^{tr}} & \mathcal{A}_0 & \xleftarrow{f_2^{tr}} & \mathcal{A}_2 \\ \| & & \| & & \| \\ J_1 \otimes I^*(B_1) & \longrightarrow & \mathcal{S} & \longleftarrow & J_2 \end{array}$$

$$(\tilde{D}_0)\qquad \begin{array}{ccccc} \tilde{\mathcal{A}}_1 & \xrightarrow{\tilde{f}_1^{tr}} & \tilde{\mathcal{A}}_0 & \xleftarrow{\tilde{f}_2^{tr}} & \tilde{\mathcal{A}}_2 \\ \| & & \| & & \| \\ J_1 \otimes (I^*(B_1) \otimes_{\tau_1} I^*(F)) & \longrightarrow & \mathcal{S} \otimes I^*(F) & \longleftarrow & J_2 \otimes I^*(F) \end{array}$$

To avoid confusions in the various differentials involved, we shall use for differentials d_i in $\tilde{\mathcal{A}}_i$, and d_F in $I^*(F)$. For any $x \in I^*(F)$ we have then

$$(11)\qquad d_2 x = d_F x ,$$

while

$$(12)\qquad d_1 x = d_F x + d_\tau x = d_F x + \Sigma b_i x_i + \Sigma b'_j x'_j$$

with x_i, $x'_j \in I^*(F)$, b_i, $b'_j \in I^*(B_1)$ and

(13) $\deg b_i = \deg s$,

(14) $\deg b'_j > 0$, $\deg b'_j \neq \deg s$.

It is clear that under $\tilde{f}$ we should have:

$(15)_1$ $\tilde{f}_1 x = x + s\delta_1 x$ for $x \in I^*(F)$ and some $\delta_1 x \in I^*(F)$,

$(15)_2$ $\tilde{f}_1 b = 0$ for $b \in I^*(B_1)$, $\deg b > 0$, $\deg b \neq \deg s$,

$(15)_3$ $\tilde{f}_1 b = |b|\, s$ for $b \in I^*(B_1)$, $\deg b = \deg s$,

and some $|b| \in \underline{k}$.

From $\tilde{f}_1(xy) = \tilde{f}_1(x)\,\tilde{f}_1(y)$ for $x,y \in I^*(F)$ we get

$(15)_4$ $\delta_1(xy) = \delta_1 x \cdot y + (-1)^{\deg s \cdot \deg x} \cdot x\delta_1 y$.

From $d_0\,\tilde{f}_1 x = \tilde{f}_1\, d_1 x$ for $x \in I^*(F)$ with $d_1 x$ given as before, we get

$(15)_5$ $(-1)^{\deg s} \cdot d_F\, \delta_1 x = \delta_1\, d_F x + \Sigma\, |b_i|\, x_i$.

Similarly, for $\tilde{f}_2$ we have

$(16)_1$ $\tilde{f}_2 x = x + s \cdot \delta_2 x$,

$(16)_2$ $\delta_2(xy) = \delta_2 x \cdot y + (-1)^{\deg s \cdot \deg x} \cdot x\delta_2 y$,

$(16)_3$ $(-1)^{\deg s} \cdot d_F\, \delta_2 x = \delta_2\, d_F x$,

in which $x,y \in I^*(F)$ and $\delta_2 x$, $\delta_2 y \in I^*(F)$ too.

By definition, $U(\tilde{f}_1^{tr}, \tilde{f}_2^{tr})$ is consisting of elements of the form

$$\alpha = (\alpha_1 , \alpha_2)$$

with $\alpha_1 \in \tilde{A}_1$, $\alpha_2 \in \tilde{A}_2$, $\tilde{f}_1^{tr}\, \alpha_1 = \tilde{f}_2^{tr}\, \alpha_2$.

Similarly, for $U(f_1^{tr}, f_2^{tr})$. It is thus clear that the following are all elements of $U(\tilde{f}_1^{tr}, \tilde{f}_2^{tr})$:

$(17)_1$ $\xi = (x,\ x + s_2(\delta_1 x - \delta_2 x))$, $x \in I^*(F)$.

$(17)_2$ $\beta' = (b', 0)$, $b' \in I^*(B_1)$, $\deg b' > 0$, $\deg b' \neq \deg s$.

$(17)_3$ $\beta = (b,\ |b|\, s_2)$, $b \in I^*(B_1)$, $\deg b = \deg s$.

$(17)_4$ $\sigma = (s_1 , s_2)$.

(17)$_5$ $\tau_1 = (t_1\ ,\ 0)$.

(17)$_6$ $\tau_2 = (0,\ t_2)$.

The elements β', β, σ, τ_1, τ_2 are in fact also elements of $U(f_1^{tr}, f_2^{tr})$ which may be considered a sub-DGA of $U(\tilde{f}_1^{tr}, \tilde{f}_2^{tr})$ by natural inclusion. From the above formulae we see that the set of all elements of type ξ with $x \in I^*(F)$ form a sub-GA of $U(\tilde{f}_1^{tr}, \tilde{f}_2^{tr})$ isomorphic as GA to $I^*(F)$ under the morphism

(18) $$h : x \to (x,\ x + s_2(\delta_1 x - \delta_2 x)) .$$

Introduce now a differential d_F in this GA by setting

(19) $$d_F(x,\ x + s_2(\delta_1 x - \delta_2 x)) = (d_F x,\ d_F x + s_2(\delta_1 d_F x - \delta_2 d_F x)) .$$

Then, this GA will become a DGA and the above morphism h will become a DGA-isomorphism.

Now, as a GA, $U(\tilde{f}_1^{tr}, \tilde{f}_2^{tr})$ is in fact the tensor product of $U(f_1^{tr}, f_2^{tr})$ and $hI^*(F)$ which may be seen as follows. We have for $x \in I^*(F)$, $b, b' \in I^*(B)$, $\deg b = \deg s$, $\deg b' > 0$, $\deg b' \neq \deg s$:

(20) $$\begin{cases} (s_1 x,\ s_2 x) = (s_1\ ,\ s_2)(x,\ x + s_2(\delta_1 x - \delta_2 x)) , \\ (bx,\ |b| s_2 x) = (b,\ |b| s_2)(x,\ x + s_2(\delta_1 x - \delta_2 x)) , \\ (b'x,\ 0) \quad = (b'\ ,\ 0)(x,\ x + s_2(\delta_1 x - \delta_2 x)) , \\ (t_1 x,\ 0) \quad = (t_1\ ,\ 0)(x,\ x + s_2(\delta_1 x - \delta_2 x)) , \\ (s_1 b\ ,\ 0) \quad = (s_1\ ,\ s_2)(b,\ |b| s_2) , \\ (s_1 b',\ 0) \quad = (s_1\ ,\ s_2)(b',0) , \\ (0,\ t_2 x) \quad = (0,\ t_2)(x,\ x + s_2(\delta_1 x - \delta_2 x)) , \quad \text{etc.} \end{cases}$$

In these formulae each right-hand side is an expression formed from elements in $U(f_1^{tr}, f_2^{tr})$ and $hI^*(F)$ by algebraic operations. Thus, it follows easily as a GA :

$$U(\tilde{f}_1^{tr}, \tilde{f}_2^{tr}) \approx U(f_1^{tr}, f_2^{tr}) \otimes hI^*(F) .$$

Consider now any element

$$\xi = (x,\ x + s_2(\delta_1 x - \delta_2 x)) \in hI^*(F) ,$$

with $d_1 x = d_F x + \Sigma b_i x_i + \Sigma b'_j x'_j$ as before. As element of $U(\tilde{f}_1^{tr}, \tilde{f}_2^{tr})$, the differential of ξ, using (15) - (20), will be given by

$$d\xi = (d_1 x, d_2 x + t_2(\delta_1 x - \delta_2 x) + (-1)^{\deg s} \cdot s_2 d_2(\delta_1 x - \delta_2 x))$$
$$= \xi_1 + \xi_2 + \xi_3 ,$$

where

$$\xi_1 = d_F \xi \in hI^*(F) ,$$

$$\xi_2 = \Sigma (b_i x_i , s_2 |b_i| x_i) + \Sigma (b'_j x'_j , 0) ,$$

$$\xi_3 = (0, t_2(\delta_1 x - \delta_2 x)) .$$

It follows that the differential d in $U(\tilde{f}_1^{tr}, \tilde{f}_2^{tr})$ consists of two parts:

$$d = d_\otimes + d_\tau .$$

The first part $d_\otimes$ is the usual one for a tensor product, while the second twisted part is given by

$$d_\tau \xi = \xi_2 + \xi_3 , \quad \xi \in hI^*(F) .$$

Hence, we may write as a DGA-isomorphism,

$$U(\tilde{f}_1^{tr}, \tilde{f}_2^{tr}) \simeq U(f_1^{tr}, f_2^{tr}) \otimes_\tau hI^*(F) .$$

Now, by theorems in § VI.2 ,

$$I^*(B) \approx \min U(f_1^{tr}, f_2^{tr}) .$$

Using the theorems of § II.6 , we can construct a DGA $I^*(B) \otimes_\tau I^*(F)$ with a certain twisted differential $d = d_\otimes + d_\tau$ as well as a DGA-morphism

$$\tilde{\rho} : I^*(B) \otimes_\tau I^*(F) \to U(f_1^{tr}, f_2^{tr}) \otimes_\tau hI^*(F)$$

which will induce an H-isomorphism. It follows that

$$\min(I^*(B) \otimes_\tau I^*(F)) \approx \min(U(f_1^{tr}, f_2^{tr}) \otimes_\tau hI^*(F))$$
$$\simeq \min U(\tilde{f}_1^{tr}, \tilde{f}_2^{tr}) \simeq I^*(E) .$$

This completes the induction and, thus, proves the theorem.

Remark: From the proof we see that the twisted part of the differential consits of two parts viz., ξ_2, inherited from the previous twisted part in the fibration E_1 over B_1, and ξ_3, a new one arisen from the adjoining of E_2 over B_2 to E_1 along E_0 with a twist. This accounts for the twisted part in the final result.

<u>Corollary</u>: For a simplicial fibration $(\mathcal{F})$ the H^*-measure of the fiber space E is given by

$$(2)' \qquad H^*(E) \simeq H(I^*(B) \underset{\tau}{\otimes} I^*(F)) .$$

<u>Remark</u>: In the formulae (2) and (2)' we cannot replace $I^*(B)$ or $I^*(F)$ by $H^*(B)$ or $H^*(F)$, only in quite exceptional cases. Thus, for the determination of $H^*(E)$ mere knowledge of $H^*(B)$, $H^*(F)$ and whatever interrelations between them are not sufficient.
On the other hand, the knowledge of $I^*(B)$, $I^*(F)$ and their interrelations expressed in terms of the twisted differential d_τ is sufficient to determine $I^*(E)$ (and hence also $H^*(E)$) completely. This is just what we mean by <u>calculability</u> of the I^*-measure in the present case.

VI.6 SOME APPLICATIONS TO FIBRATIONS - TRANSGRESSION AND A THEOREM OF BOREL-HIRSCH

As a further application of the method developed in this chapter let us consider a theorem of Borel-Hirsch for certain kind of fibrations. This theorem is of basic importance for the whole theory of the I^*-measure, as it can be seen from § IV.2. To begin with, let us consider a simplicial fibration

$$(\mathcal{F}) \qquad F \underset{i}{\subset} E \xrightarrow[g]{} B$$

and the associated diagram below:

$$(D_A) \qquad A^*(F) \xleftarrow[i^A]{} A^*(E) \xleftarrow[g^A]{} A^*(B) .$$

With a well-known definition as a prototype we now lay down the following

<u>Definition</u>: A cycle $u \in A^*(F)$ will be said to be <u>transgressive</u> in the fibration $(\mathcal{F})$ if there exist some element $a \in A^*(E)$ and a cycle $x \in A^*(B)$ such that

$$i^A a = u \quad \text{and} \quad g^A x = da .$$

In this case x will then be called a <u>transgression</u> of the transgressive element u.

<u>Lemma</u>: If a cycle $u \in A^*(F)$ is transgressive with a cycle $x \in A^*(B)$ as its transgression, then any cycle $u' \in A^*(F)$ homologous to u is also transgressive, and any cycle $x' \in A^*(B)$ homologous to x is a transgression of u'.

<u>Proof</u>: Let $a \in A^*(E)$ be as above. Let $u' = u + dv$, $x' = x + dy$ for some $v \in A^*(F)$, $y \in A^*(B)$. Since F is a subcomplex of E so that i^A is onto by § VI.1, there will be some element $b \in A^*(E)$ with $i^A b = v$. Set $a' = a + g^A y + db$. Then, we have

$$da' = da + g^A dy = g^A x',$$

and

$$i^A a' = i^A a + i^A db = u + dv = u',$$

since $i^A g^A y$ clearly is 0. This shows, that u' is transgressive with x' a transgression, as to be proved.

From the lemma we see that it is legitimate to lay down the following

<u>Definition</u>: A class U of $H^*(F)$ is said to be <u>transgressive</u> with a class X of $H^*(B)$ as a transgression, if there are cycles u of $A^*(F)$ and x of $A^*(B)$ belonging to U and X respectively when passing to homology, such that u is transgressive with x as a transgression.

Consider now a simplicial fibration $(\mathcal{F})$ verifying the following conditions:

(C_1) $I^*(F)$ is free, of finite type, and has a trivial differential.

(C_2) Each free generator of $I^*(F) \approx H^*(F)$ is transgressive.

<u>Definition</u>: A simplicial fibration $(\mathcal{F})$ verifying conditions (C_1) and (C_2) will be said to be <u>totally transgressive</u>.

For such a totally transgressive fibration we have $I^*(F) \approx H^*(F) \approx \text{Free}(\tilde{u}_j)$ with all $d\tilde{u}_j = 0$, say. For some associated minimal morphism $\rho_F : I^*(F) \to A^*(F)$ let $u_j = \rho_F \tilde{u}_j$. Then, each u_j is transgressive, and we may choose for u_j a transgression $x_j \in A^*(B)$. Let $\rho_B : I^*(B) \to A^*(B)$ be some associated minimal morphism. By the Lemma we may choose some cycle $\tilde{x}_j$ in $I^*(B)$ without loss of generality, such that $\rho_B \tilde{x}_j = x_j$ for each x_j. Form now a DGA

$$\mathcal{E} = I^*(B) \underset{\tau}{\otimes} I^*(F)$$

with a twisted differential

(1) $$d = d_{\otimes} + d_{\tau}$$

of which the twisted part d_τ is given by

(2) $$d_\tau \tilde{u}_j = \tilde{x}_j .$$

With the well-known theorem of Borel-Hirsch as a prototype we then have the following

<u>Theorem</u>: For a totally transgressive simplicial fibration (Γ) the I^*-measure of the fiber space E is given by

(3) $$I^*(E) \approx \min(I^*(B) \underset{\tau}{\otimes} I^*(F))$$

with a twisted differential defined by (1) - (2).

<u>Proof</u>: For $u_j = \rho_F \tilde{u}_j$, and $x_j = \rho_B \tilde{x}_j$ as above, u_j is transgressive with a transgression x_j in the diagram (D_A), so that we have for some $a_j \in A^*(E)$:

(4) $$i^A a_j = u_j , \quad g^A x_j = da_j .$$

Form the DGA $\mathcal{E}_A = A^*(B) \underset{\tau}{\otimes} I^*(F)$ with the twisted part of a twisted differential in the tensor product given by

(5) $$d_\tau \tilde{u}_j = x_j .$$

Then, we see from known theorems of Moore in § II.6 that the natural DGA-morphism

$$\rho_B \otimes \text{ident.} : I^*(B) \underset{\tau}{\otimes} I^*(F) \to A^*(B) \underset{\tau}{\otimes} I^*(F)$$

is an H-isomorphism so that

(6) $$\min(I^*(B) \underset{\tau}{\otimes} I^*(F)) \approx \min(A^*(B) \underset{\tau}{\otimes} I^*(F)) .$$

It follows that it will be sufficient to prove

(7) $$I^*(E) \approx \min(A^*(B) \underset{\tau}{\otimes} I^*(F)) .$$

For this we shall restrict ourselves to the case of B being a finite connected complex and proceed by induction as in the last section.

Suppose therefore B be the union of two connected subcomplexes B_1, B_2 with a connected subcomplex B_0 in common. Let the parts of fibration $(\mathcal{F})$ over B_s $(s=0,1,2)$ be

$$(\mathcal{F}_s) \qquad F \underset{i_s}{\subset} E_s \xrightarrow[g_s]{} B_s .$$

For $s=1,2$ let $f_s : B_0 \subset B_s$, $\tilde{f}_s : E_0 \subset E_s$ be inclusions. Similarly, let $k_s : B_s \subset B$, $\tilde{k}_s : E_s \subset E$ also be inclusions for each $s=0,1,2$. From (4) we get then

$$i_s^A \tilde{k}_s^A a_j = u_j , \qquad g_s^A k_s^A x_j = d\tilde{k}_s^A a_j .$$

Hence, each fibration $(\mathcal{F}_s)$ is totally transgressive, with $k_s^A x_j$ being a transgression of the cycle $u_j \in A^*(F)$. We shall suppose that the theorem has been proved for each fibration $(\mathcal{F}_s)$ with

$$(\mathcal{F}_s) \qquad I^*(E_s) \approx \min\,(A^*(B_s) \underset{\tau_s}{\otimes} I^*(F)),$$

in which the twisted part of the twisted differential is defined by

$$d_{\tau_s} \tilde{u}_j = k_s^A x_j .$$

Proceed now to the proof of (7) as follows. Let us set $(s=0,1,2$ or $s=1,2)$

$$\begin{aligned}
\bar{A}^*(B_s) &= A^*(B_s) \underset{\tau_s}{\otimes} I^*(F) , \\
\bar{I}^*(E_s) &= I^*(E_0)^{tr} \otimes I^*(E_s) , \\
\bar{f}_s^A &= f_s^A \otimes \text{id.} : \bar{A}^*(B_s) \to \bar{A}^*(B_0) , \\
\bar{f}_s^I &= tr \otimes \tilde{f}_s^I \ : \bar{I}^*(E_s) \to I^*(E_0) , \\
\tilde{f}_s^I & \qquad\qquad : I^*(E_s) \to I^*(E_0) .
\end{aligned}$$

in which

$$tr : I^*(E_0)^{tr} \to I^*(E_0)$$

is the trivializing morphism for $I^*(E_0)$.

Since the DGA-morphisms f_1^A, f_2^A in the diagram below

$$A^*(B_1) \xrightarrow{f_1^A} A^*(B_0) \xleftarrow{f_2^A} A^*(B_2)$$

both are onto, so that $\bar{f}_s^A$ are also onto, we can complete a <u>commutative</u> diagram of DGA-morphisms as shown below, with ρ_s some minimal morphisms and $\bar{\rho}_s = \varepsilon_0 \otimes \rho_s$ with $\varepsilon_0 : I^*(E_0)^{tr} \to \underline{k}$ the augmentation:

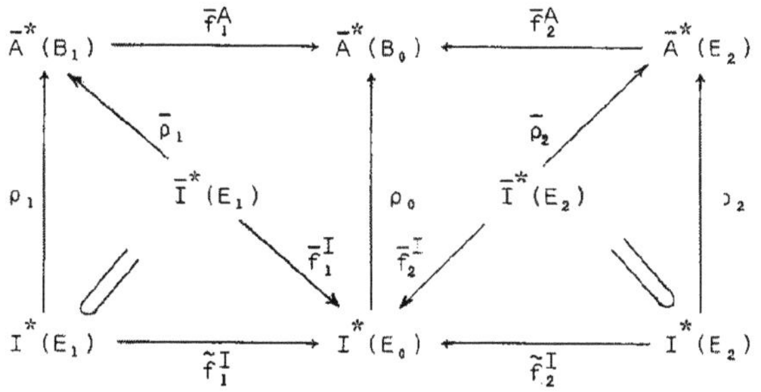

From theorems of § VI.2 we have now

$$I^*(E) \approx \min U(\bar{f}_1^I, \bar{f}_2^I)$$
$$\approx \min U(\bar{f}_1^A, \bar{f}_2^A) .$$

It is clear that

$$U(\bar{f}_1^A, \bar{f}_2^A) \approx U(f_1^A, f_2^A) \otimes_\tau I^*(F) \approx A^*(B) \otimes_\tau I^*(F)$$

with the twisted part of the twisted differential or the right-hand side, given by

$$d_\tau \tilde{u}_j = (d_{\tau_1} \tilde{u}_j , d_{\tau_2} \tilde{u}_j) = (k_1^A x_j , k_2^A x_j) = x_j .$$

Whence we get (7) which completes the induction and proves the theorem.

<u>Corollary</u>: For a totally transgressive simplicial fibration (F) the H^*-measure of the fiber space E is given by

$$H^*(E) \approx H(I^*(B) \otimes_\tau H^*(F)) ,$$

in which the twisted differential is determined by the transgressions.

<u>Remark</u>: In the theorem of the last section the twisted part d_τ of the twisted differential is only shown to exist but no constructive means has been indicated. However, in the totally transgressive case of the theorem in this section the twisted part is constructively and explicitely given via the transgression. In the next chapter, further particular cases will be given, for which the twisted differential can explicitely be determined.

Chapter VII

I^*-MEASURES CONNECTED WITH FIBRATIONS

VII.1 SOME ALGEBRAIC PREPARATIONS

In this section we shall make some algebraic preparations to be used in the later sections for the study of the I^*-measures connected with various kinds of fibrations.

Definition: A DGA $\mathcal{A}$ will be said to be a right- (resp. left-) DGA over a DGA $\mathcal{B}$ or a right- (resp. left-) $\mathcal{B}$-DGA if there is a DGA-morphism

$$\gamma : \mathcal{A} \otimes \mathcal{B} \to \mathcal{A}$$

$$(\gamma' : \mathcal{B} \otimes \mathcal{A} \to \mathcal{A} \quad \text{resp.})$$

We shall also say that $\mathcal{B}$ operates on the right (resp. left) on $\mathcal{A}$ and, if no ambiguity can occur, write ($\alpha \in \mathcal{A}$, $\beta \in \mathcal{B}$):

$$\gamma(\alpha \otimes \beta) = \alpha\beta ,$$

$$(\gamma'(\beta \otimes \alpha) = \beta\alpha \quad \text{resp.}) .$$

Suppose $\mathcal{A}'$ be a right-DGA and $\mathcal{A}''$ be a left-DGA over the same DGA $\mathcal{B}$. In the tensor product $\mathcal{A}' \otimes \mathcal{A}''$ consider the ideal I generated by elements of the form

$$a'b \otimes a'' - a' \otimes ba''$$

with $a' \in \mathcal{A}'$, $a'' \in \mathcal{A}''$, $b \in \mathcal{B}$. As the differential of an element in I is easily seen to be still an element of I, so the gradation and differentiation in $\mathcal{A}' \otimes \mathcal{A}''$ will also induce ones in the quotient algebra $\mathcal{A}' \otimes \mathcal{A}''/I$ to turn it into a DGA. Now, we use the following

Notation: The DGA $\mathcal{A}' \otimes \mathcal{A}''/I$ with gradation and differentiation induced from those of $\mathcal{A}' \otimes \mathcal{A}''$ will be denoted by $\mathcal{A}' \otimes_{\mathcal{B}} \mathcal{A}''$.

Definition: Give a DGA-morphism

$$f : \mathcal{B} \to \mathcal{A}$$

of DGA $\mathcal{B}$ to DGA $\mathcal{A}$,

$$\gamma(\alpha \otimes \beta) = \alpha \cdot f(\beta) \quad (\text{or simply } \alpha\beta)$$

for $\alpha \in \mathcal{A}$, $\beta \in \mathcal{B}$ will naturally turn $\mathcal{A}$ into a right-DGA over $\mathcal{B}$, which wil be said to be <u>induced</u> from f and will be denoted by

$$\gamma = R(f) .$$

Similarly,

$$\gamma'(\beta \otimes \alpha) = f(\beta)\cdot\alpha \quad (\text{or simply } \beta\alpha)$$

will turn $\mathcal{A}$ into a left-DGA over $\mathcal{B}$ <u>induced</u> by f to be denoted by

$$\gamma' = L(f) .$$

Given a right-DGA $\mathcal{A}$ over DGA $\mathcal{B}$ with operation induced by DGA-morphism $f : \mathcal{B} \to \mathcal{A}$ let us put

$$\Omega_n \mathcal{B} = \underbrace{\mathcal{B}^+ \otimes \ldots \otimes \mathcal{B}^+}_{n} ,$$

$$\Omega \mathcal{B} = \sum_{n\geq 0} \Omega_n \mathcal{B} ,$$

$$\mathrm{Bar}_n^{R(f)}(\mathcal{A}) = \mathcal{A} \otimes \Omega_n \mathcal{B} \otimes \mathcal{B} ,$$

$$\mathrm{Bar}^{R(f)}(\mathcal{A}) = \sum_{n\geq 0} \mathrm{Bar}_n^{R(f)}(\mathcal{A}) = \sum_{n\geq 0} \mathcal{A} \otimes \Omega_n(\mathcal{B}) \otimes \mathcal{B} .$$

In case $n = 0$, $\Omega_0\mathcal{B}$ will be considered generated by a single element [] of degree 0. An element $\xi = a \otimes b_1 \otimes \ldots \otimes b_n \otimes b$ in $\mathrm{Bar}^{R(f)}(\mathcal{A})$ with $a \in \mathcal{A}$, $b_i \in \mathcal{B}^+$, $b \in \mathcal{B}$, will be written in the form $\xi = a[b_1|\ldots|b_n]b$, as usual. Now, introduce a gradation deg and differentiation d in $\mathrm{Bar}^{R(f)}(\mathcal{A})$ by setting

$$\deg \xi = \deg a + \deg b + \sum_{i=1}^{n} (\deg b_i - 1) ,$$

$$d = d_E + d_I ,$$

$$\begin{aligned} d_E \xi &= (-1)^{d_0}\cdot ab_1[b_2|\ldots|b_n]b \\ &\quad + \sum_{i=1}^{n-1} (-1)^{d_i}\cdot a[b_1|\ldots|b_i b_{i+1}|\ldots b_n]b \\ &\quad + (-1)^{d_n}\cdot a[b_1|\ldots|b_{n-1}]\, b_n b , \end{aligned}$$

$$\begin{aligned} d_I \xi &= da\cdot[b_1|\ldots|b_n]b - \sum (-1)^{d_{i-1}}\cdot a[b_1|\ldots|db_i|\ldots|b_n]b \\ &\quad + (-1)^{d_n}\cdot a[b_1|\ldots|b_n]\cdot db , \end{aligned}$$

in which $ab_1 = af(b_1)$, and

$$d_0 = \deg a, \quad d_i = \deg a + \sum_{j=1}^{i} (\deg b_j - 1), \quad i = 1, \ldots, n.$$

It is easy to verify that

$$d_E^2 = 0, \quad d_I^2 = 0,$$

$$d_I d_E + d_E d_I = 0,$$

so that $d^2 = 0$, or d really is a differential in $\mathrm{Bar}^{R(f)}(\mathcal{A})$.

Using the usual shuffle multiplication which we shall not explicit here, $\Omega(\mathcal{B})$ and hence also $\mathrm{Bar}^{R(f)}(\mathcal{A})$, will become a DGA with the above-defined gradation and differentiation. Now, we lay down the following

Definition: The DGA $\mathrm{Bar}^{R(f)}(\mathcal{A})$ will be called the Bar-Construction of $\mathcal{A}$ as right-DGA over $\mathcal{B}$ induced by the DGA-morphism $f : \mathcal{B} \to \mathcal{A}$.

Suppose now, we have DGA's $\mathcal{A}', \mathcal{A}'', \mathcal{B}$ and DGA-morphisms

$$f' : \mathcal{B} \to \mathcal{A}',$$

$$f'' : \mathcal{B} \to \mathcal{A}''$$

which turn $\mathcal{A}'$ and $\mathcal{A}''$ into right- and left-DGA's over $\mathcal{B}$ respectively. Since it is clear that $\mathrm{Bar}^{R(f)}(\mathcal{A}')$ will also be a right-DGA over $\mathcal{B}$, induced from the natural inclusion morphism $\mathcal{B} \to \mathrm{Bar}^{R(f')}(\mathcal{A}')$, we can form the quotient DGA

$$\mathrm{Bar}^{R(f')}(\mathcal{A}') \underset{\mathcal{B}}{\otimes} \mathcal{A}'' = \sum_{n \geq 0} (\mathcal{A}' \otimes \Omega_n(\mathcal{B}) \otimes \mathcal{B}) \underset{\mathcal{B}}{\otimes} \mathcal{A}'' = \sum_{n \geq 0} (\mathcal{A}' \otimes \Omega_n(\mathcal{B}) \otimes \mathcal{A}'').$$

Notation: The above quotient DGA will be denoted by

$$\mathrm{Bar}(f', f'') = \mathrm{Bar}^{R(f')}(\mathcal{A}') \underset{\mathcal{B}}{\otimes} \mathcal{A}''.$$

In particular, if $\mathcal{A}'' = \mathcal{B}$ and f'' is the identity $\iota_{\mathcal{B}} : \mathcal{B} \to \mathcal{A}''$. Then

$$\mathrm{Bar}(f', \iota_{\mathcal{B}}) = \mathrm{Bar}^{R(f')}(\mathcal{A}').$$

Definition: The homology of the above quotient DGA is called the Tor-product of the right- and left-DGA's $\mathcal{A}', \mathcal{A}''$ over $\mathcal{B}$ under f', f'', to be denoted by

$$\mathrm{Tor}(f', f'') = H(\mathrm{Bar}(f', f'')),$$

which is $\mathrm{Tor}_{\mathcal{B}}(\mathcal{A}', \mathcal{A}'')$ in the current notations.

Proposition 1: If $\mathcal{A}' = \mathcal{B}$ with $f' : \mathcal{B} \to \mathcal{A}'$ the identity morphism $\iota_{\mathcal{B}}$, while $\mathcal{A}''$ is the basic field $\underline{k}$ and $f'' : \mathcal{B} \to \underline{k}$ is the augmentation $\varepsilon_{\mathcal{B}}$, then $\mathrm{Bar}\,(\iota_{\mathcal{B}}, \varepsilon_{\mathcal{B}})$ is homologically trivial, or

$$\mathrm{Tor}\,(\iota_{\mathcal{B}}, \varepsilon_{\mathcal{B}}) \approx \underline{k}\ .$$

If $\mathcal{A}'$, f' remain as above while $\mathcal{A}'' = \mathcal{B}$ with f'' the identity morphism $\iota_{\mathcal{B}}$ too, then

$$\mathrm{Tor}\,(\iota_{\mathcal{B}}, \iota_{\mathcal{B}}) \approx H\,(\mathcal{B})\ .$$

Proof: Define a module-morphism s of $\mathrm{Bar}\,(\iota_{\mathcal{B}}, \varepsilon_{\mathcal{B}})$ lowering degree 1 by $s\,(b[b_1|\dots|b_n]) = [b|b_1|\dots|b_n]$.

Then, we have by direct verification

$$ds + sd = \mathrm{ident}.$$

From this follows the homological-triviality of $\mathrm{Bar}\,(\iota_{\mathcal{B}}, \varepsilon_{\mathcal{B}})$ and, whence, also the other assertion.

Suppose there is a commutative diagram of DGA's and DGA-morphisms as shown below:

$$\begin{array}{ccccc} \mathcal{A}' & \xleftarrow{f'} & \mathcal{B} & \xrightarrow{f''} & \mathcal{A}'' \\ {\scriptstyle g'}\uparrow & & \uparrow{\scriptstyle g} & & \uparrow{\scriptstyle g''} \\ \bar{\mathcal{A}}' & \xleftarrow{\bar{f}'} & \bar{\mathcal{B}} & \xrightarrow{\bar{f}''} & \bar{\mathcal{A}}'' \end{array}$$

The triple (g', g'', g) will then induce in a natural manner a DGA-morphism

$$\tilde{g} : \mathrm{Bar}\,(\bar{f}', \bar{f}'') \to \mathrm{Bar}\,(f', f'')$$

defined by

$$\tilde{g}\,(\bar{a}'[\bar{b}_1|\dots|\bar{b}_n]\bar{a}'') = g'\,\bar{a}'[g\bar{b}_1|\dots|g\bar{b}_n]\cdot g''\,\bar{a}''$$

with $\bar{a}' \in \bar{\mathcal{A}}'$, $\bar{a}'' \in \bar{\mathcal{A}}''$, $\bar{b} \in \bar{\mathcal{B}}^+$.

Proposition 2: If in the above commutative diagram g', g'' and g all are H-isomorphisms then $\tilde{g}$ is also an H-isomorphism, so that

$$\tilde{g}_H : \mathrm{Tor}\,(\bar{f}', \bar{f}'') \approx \mathrm{Tor}\,(f', f'')\ .$$

<u>Proof</u>: Omitted.

We have now the following theorem of Eilenberg-Moore:

<u>Theorem 1</u>: Let $f' : \mathcal{B} \to \mathcal{A}'$, $f'' : \mathcal{B} \to \mathcal{A}''$ be DGA-morphisms turning $\mathcal{A}', \mathcal{A}''$ into right- resp. left-DGA's over $\mathcal{B}$, as before. Then, there is a filtration in Bar (f', f") with

$$E_2 \approx \mathrm{Tor}\,(f'_H, f''_H) \Longrightarrow \mathrm{Tor}\,(f', f'')$$

in which $f'_H : H(\mathcal{B}) \to H(\mathcal{A}')$, $f''_H : H(\mathcal{B}) \to H(\mathcal{A}'')$ are induced by f', f'', turning $H(\mathcal{A}')$, $H(\mathcal{A}'')$ into right- and left-DGA's over $H(\mathcal{B})$ respectively.

<u>Definition</u>: The spectral sequence of the above theorem is called the <u>Eilenberg-Moore spectral sequence</u>, associated to the DGA-morphisms f' and f".

Consider now the special case for which $\mathcal{B}$ is a polynomial DGA.

$$\mathcal{B} \approx \mathrm{Polym}\,(y_1, \dots, y_N),$$

$$\deg y_i = \text{even} > 0, \quad dy_i = 0.$$

In this case, some much simpler constructions, due to Koszul, may then replace the Bar-construction for the definition of the Tor-product.

To describe it, let $\mathcal{A}$ be a DGA and

$$f : \mathcal{B} \to \mathcal{A}$$

be a DGA-morphism turning $\mathcal{A}$ into a right-DGA over the polynomial DGA $\mathcal{B}$. Form the tensor product

$$\mathrm{Kosz}^{R(f)}(\mathcal{A}) = \mathcal{A} \otimes \mathrm{Extr}\,(x_1, \dots, x_N) \otimes \mathcal{B}$$

with degree and differential d defined by

$$\deg x_i = \deg y_i - 1 \quad \text{odd},$$

$$d = d_E + d_I,$$

$$d_E\,(a \otimes 1 \otimes b) = 0,$$

$$d_E\,(a \otimes x_i \otimes b) = (-1)^{\deg a} \cdot ay_i \otimes 1 \otimes b + (-1)^{\deg a + \deg x_i} \cdot a \otimes 1 \otimes y_i b,$$

$$d_I\,(a \otimes 1 \otimes b) - da \otimes 1 \otimes b + (-1^{\deg a} \cdot a \otimes 1 \otimes db$$

$$d_I\,(a \otimes x_i \otimes b) = da \otimes x_i \otimes b + (-1)^{\deg a + \deg x_i} \cdot a \otimes x_i \otimes db$$

in which $a \in \mathcal{A}$, $b \in \mathcal{B}$. Then, we have $d_E^2 = 0$, $d_I^2 = 0$, $d_I d_E + d_E d_I = 0$ so that $d = d_E + d_I$ can be extended to a differential d on $\mathrm{Kosz}^{R(f)}(\mathcal{A})$.

Definition: The DGA $\mathrm{Kosz}^{R(f)}(\mathcal{A})$ will be called the Koszul-Construction of $\mathcal{A}$ as right-DGA over the polynomial DGA $\mathcal{B}$, induced by the DGA-morphism $f : \mathcal{B} \to \mathcal{A}$.

Notation: Let $f' : \mathcal{B} \to \mathcal{A}'$, $f'' : \mathcal{B} \to \mathcal{A}''$ be DGA-morphisms turning $\mathcal{A}'$, $\mathcal{A}''$ into right- and left-DGA's over the polynomial DGA $\mathcal{B}$ respectively. Then, we shall denote the quotient DGA

$$\mathrm{Kosz}^{R(f')}(\mathcal{A}') \underset{\mathcal{B}}{\otimes} \mathcal{A}'' = \mathcal{A}' \otimes \mathrm{Extr}(x_1, \ldots, x_N) \otimes \mathcal{A}''$$

by $\mathrm{Kosz}(f', f'')$.

Theorem 2: Let $\mathcal{B} \approx \mathrm{Polym}(y_1, \ldots, y_N)$ be a polynomial DGA and $\mathcal{A}'$, $\mathcal{A}''$ be right- and left-DGA's over $\mathcal{B}$, induced by DGA-morphisms $f' : \mathcal{B} \to \mathcal{A}'$ and $f'' : \mathcal{B} \to \mathcal{A}''$ respectively. Then, $x_i \to [y_i]$ will induce a DGA-morphism h of $\mathrm{Kosz}(f', f'')$ into $\mathrm{Bar}(f', f'')$ which is an H-isomorphism, so that

$$\mathrm{Tor}(f', f'') \approx H(\mathrm{Kosz}(f', f'')).$$

Proof: That $\mathrm{Kosz}(f', f'')$ and $\mathrm{Bar}(f', f'')$ have isomorphic homology follows from a general theory of resolutions which we shall not enter.

Now, define the morphism h of $\mathrm{Kosz}(f', f'')$ into $\mathrm{Bar}(f', f'')$ by

$$h(a \otimes 1 \otimes b) = a[\,]b,$$

$$h(a \otimes x_{i_1}, \ldots, x_{i_p} \otimes b) = \sum_s \varepsilon_s [y_{s(i_1)} | \ldots | y_{s(i_p)}]$$

in which the summation is over all permutations s of $(i_1, \ldots, i_p)$ and $\varepsilon_s = +1$ or -1 according to the permutation s being even or odd. It is easy to see that multiplication in x_i goes to shuffle multiplication in $[y_i]$ under h, so the morphism h is multiplicative. It also preserves gradation as well as differentiation, as seen from the respective formulae for differentials. Hence, h is a DGA-morphism inducing isomorphism of homology or an H-isomorphism, as to be proved.

VII.2 SIMPLICIAL FIBRATION AND SOME SPECTRAL SEQUENCES

Consider a fibration

$$(\mathcal{F}) \qquad F \underset{i}{\subset} Y \xrightarrow[g]{} B$$

in which F is the fiber, Y the fiber space, and B is the base space. We shall suppose that the fibration is simplicial in the sense of § VI.5 - 6 .

Let B_q be the q-skeleton of B, $Y_q = g^{-1}(B_q)$ and $i_q : Y_q \subset Y$ be the inclusion. Now, define a filtration of $A^*(Y)$ by

$$(1)_A \qquad F^p A^q(Y) = \mathrm{Ker}\,[A^q(Y) \xrightarrow{i^A_{p-1}} A^q(Y_{p-1})]$$

so that

$$(2)_A \qquad A^*(Y) = F^0 \supset F^1 \supset \ldots \supset F^p \supset F^{p+1} \supset \ldots .$$

Lemma 1: For the Leray spectral sequence of filtration $(1)_A$ of $A^*(Y)$ we have

$$(3) \qquad E_2^{p,q} A^*(Y) \approx H^p(B) \otimes H^q(F) .$$

Proof: Let us introduce a filtration of $C^*(Y)$ by

$$(1)_C \qquad F^p C^q(Y) = \mathrm{Ker}\,[C^q(Y) \xrightarrow{i^C_{p-1}} C^q(Y_{p-1})]$$

so that

$$(2)_C \qquad C^*(Y) = F^0 \supset F^1 \supset \ldots \supset F^p \supset F^{p+1} \supset \ldots .$$

According to § III.4 , there is an integration

$$\int : A^*(Y) \to C^*(Y)$$

which is a DGA morphism inducing isomorphism of H or an H-isomorphism. It is clear that

$$\int F^p A^q(Y) \subset F^p C^q(Y)$$

so that $\int$ will induce morphisms of corresponding spectral sequences

$$\int_r : E_r^{p,q} A^*(Y) \to E_r^{p,q} C^*(Y) .$$

We now have a commutative diagram

$$
(D)_A \qquad
\begin{array}{ccccccccc}
 & & 0 & & 0 & & & & \\
 & & \downarrow & & \downarrow & & & & \\
0 & \longrightarrow & F^{p+1} A^{p+q}(Y) & = & F^{p+1} A^{p+q}(Y) & & & & \\
 & & \downarrow & & \downarrow & & & & \\
0 & \longrightarrow & F^{p} A^{p+q}(Y) & \longrightarrow & A^{p+q}(Y) & \longrightarrow & A^{p+q}(Y_{p-1}) & \longrightarrow & 0 \\
 & & \downarrow & & \downarrow & & \| & & \\
0 & \longrightarrow & A^{p+q}(Y_p, Y_{p-1}) & \longrightarrow & A^{p+q}(Y_p) & \longrightarrow & A^{p+q}(Y_{p-1}) & \longrightarrow & 0 \\
 & & \downarrow & & \downarrow & & & & \\
 & & 0 & & 0 & & & &
\end{array}
$$

in which all rows and columns are exact sequences, and $A^{p+q}(Y_p, Y_{p-1})$ is introduced and defined by the exactness of the last row. The differential d in $A^*(Y_p)$ and $A^*(Y_{p-1})$ will induce one in $A^*(Y_p, Y_{p-1}) = \Sigma A^r(Y_p, Y_{p-1})$ to make it a differential graded module.

From the diagram $(D)_A$ we get

$$E_0^{p,q} A^*(Y) \approx F^p A^{p+q}(Y) / F^{p+1} A^{p+q}(Y)$$

$$\approx A^{p+q}(Y_p, Y_{p-1})$$

with the differential d_0 in E_0 as that d in $A^*(Y_p, Y_{p-1})$. Hence,

$$E_1^{p,q} A^*(Y) \approx H_{p+q} A^*(Y_p, Y_{p-1}) .$$

We may form a diagram $(D)_C$ similar to $(D)_A$, by using C instead of A. With $C^*(Y_p, Y_{p-1})$ similarly defined we get then for the spectral sequence $(2)_C$

$$E_1^{p,q} C^*(Y) \approx H_{p+q} C^*(Y_p, Y_{p-1}) .$$

Since the integration morphism $\int$ is clearly functorial and $\int : A^*(Y_p) \to C^*(Y_p)$ is an (additive) H-isomorphism for all p, it will also induce an (additive) H-isomorphism

$$A^*(Y_p, Y_{p-1}) \to C^*(Y_p, Y_{p-1}) .$$

Hence

$$\int_1 : E_1^{p,q} A^*(Y) \approx E_1^{p,q} C^*(Y) .$$

From isomorphism theorems of spectral sequences we have therefore

$$\int_r : E_r^{p,q} A^*(Y) \approx E_r^{p,q} C^*(Y) , \quad r \geq 1 .$$

As it is known that $E_2^{p,q} C^*(Y) \approx H^*(B) \otimes H^*(F)$, the same will be true for $E_2^{p,q} A^*(Y)$.

This proves the Lemma.

Introduce now a further DGA and a filtration in the following way. For each vertex v_i of B let B_i be the subcomplex of B consisting of all simplexes incident with v_i as well as all their faces. Consider now a function η on the set of vertices $V = \{v_i\}$ of B with values

$$\eta(v_i) \in A^r(B_i) \otimes A^s(g^{-1} B_i)$$

such that the following compatibility condition (C) is satisfied:

For any pair of vertices v_i, v_j spanning a 1-simplex σ_{ij} of B let B_{ij} be the subcomplex of B consisting of all simplexes with σ_{ij} as face as well as all their faces. Then, $\eta(v_i)$, $\eta(v_j)$ have the same natural restrictions in $A^r(B_{ij}) \otimes A^s(g^{-1} B_{ij})$.

Denote the $\underline{k}$-module naturally formed of all such compatible collections of functions $\eta = \{\eta(v_i), \ v_i \in V/(C)\}$ by $\bar{A}^{r,s}(Y)$. Set

$$\bar{A}^q(Y) = \sum_{r+s=q} \bar{A}^{r,s}(Y) ,$$

$$\bar{A}^*(Y) = \sum_{q \geq 0} \bar{A}^q(Y) .$$

Under usual multiplications and differentiations for differential forms $\bar{A}^*(Y)$ forms naturally a DGA-algebra. Filtrate now $\bar{A}^*(Y)$ by

$$F^p \bar{A}^*(Y) = \sum_{r \geq p} \bar{A}^{r,*}(Y) . \tag{4}$$

Lemma 2: For the Leray spectral sequence of filtration (4) of $\bar{A}^*(Y)$ we have

$$E_2^p \bar{A}^*(Y) \approx H^p(B) \otimes H^*(F) . \tag{5}$$

Proof: By direct calculation we get

$$E_0^p \bar{A}^*(Y) \approx \bar{A}^{p,*}(Y)$$

$$\approx \{\eta(v_i) \in A^p(B_i) \otimes A^*(g^{-1} B_i), \quad v_i \in V/(C_0)\},$$

in which (C_0) is some natural compatibility condition, as above. The differential d_0 in $E_0^p \bar{A}^*(Y)$ is given by

$$d_0 = d \mid A^*(g^{-1} B_i).$$

Since

$$E_1^p \bar{A}^*(Y) \approx H_p(E_0 \bar{A}^*(Y)),$$

there will be a natural morphism

(6) $$E_1 \bar{A}^*(Y) \to H$$

where H is the module formed of compatible collections $\tilde{\eta}$ of functions on a set of vertices V such that

$$\tilde{\eta}(v_i) \in A^p(B_i) \otimes H^*(g^{-1} B_i), \quad v_i \in V,$$

verifying again some natural compatibility conditions $(\tilde{C})$.

We shall prove that (6) is in fact an isomorphism. Admitting this provisionally, then, as $\pi_1(B)$ acts trivially on the homology of fibers, and the fibration over each B_i is trivial, the isomorphism (6) is equivalent to say

$$E_1^p \bar{A}^*(Y) \approx A^p(B) \otimes H^*(F)$$

with differential d_1 in $E_1^p \bar{A}^*(Y)$ given by

$$d_1 = d \mid A^*(B).$$

Since $E_2 = H(E_1)$, we get the isomorphism (5) as required.

It remains to prove that (6) is an isomorphism as modules. For this purpose let us set

$$E_0 = E_0 \bar{A}^*(Y)$$

with differential d_0, and $Z(E_0)$ and $B(E_0)$ the graded sub-module of cycles and boundaries in E_0. Then, we have to prove the following

Lemma 3: The sequence of natural module-morphisms

$$0 \to B(E_0) \xrightarrow{i} Z(E_0) \xrightarrow{\pi} H \to 0$$

is exact.

Proof: Let us prove e.g. the epimorphism of π and thus consider any element $\tilde{\eta} \in H$ with $\tilde{\eta}(v_i) \in A^*(B_i) \otimes H^*(g^{-1} B_i)$, $v_i \in V$.
For each v_i let us take an arbitrary element

$$\zeta_i \in A^*(B_i) \otimes Z(A^*(g^{-1} B_i))$$

which goes to $\tilde{\eta}(v_i)$ in passing to homology. The compability condition $(\tilde{C})$ shows that for each pair of vertices v_i, v_j spanning a 1-simplex of B, there is an element

$$\varphi_{ij} \in A^*(B_{ij}) \otimes A^*(g^{-1} B_{ij})$$

such that

$$\text{Rest}_{ij} \zeta_i - \text{Rest}_{ij} \zeta_j = d_0 \varphi_{ij},$$

in which Rest_{ij} means the restriction of the corresponding element to one in $A^*(B_{ij}) \otimes A^*(g^{-1} B_{ij})$.

To φ_{ij} let us construct an element $\psi_{ij} \in A^*(B_i) \otimes A^*(g^{-1} B_i)$, as follows. Let t_j be the barycentric coordinate function in B corresponding to the vertex v_j. Write φ_{ij} in the form $\varphi_{ij} = \Sigma \varphi'_{ij\alpha} \otimes \varphi''_{ij\alpha}$ with $\varphi'_{ij\alpha} \in A^*(B_{ij})$, $\varphi''_{ij\alpha} \in A^*(g^{-1} B_{ij})$. Then, $t_j \varphi'_{ij\alpha}$ on simplexes of B_{ij} and 0 on simplexes in B_i, but not in B_{ij}, is a well-defined element in $A^*(B_i)$ which will still be denoted by $t_j \varphi'_{ij\alpha}$. By the Extension Theorem of § VI.1, we can also extend $\varphi''_{ij\alpha}$ to an element in $A^*(g^{-1} B_i)$, and we shall still denote such an arbitrary one by $\varphi''_{ij\alpha}$.

As a definition we put now

$$\psi_{ij} = \Sigma\, t_j\, \varphi'_{ij\alpha} \otimes \varphi''_{ij\alpha} \in A^*(B_i) \otimes A^*(g^{-1} B_i)$$

and write for simplicity

$$\psi_{ij} = t_j\, \varphi_{ij} .$$

Then we have

$$d_0\, \psi_{ij} = \Sigma\, t_j\, \varphi'_{ij\alpha} \otimes d\varphi''_{ij\alpha} = t_j\, d_0\, \varphi_{ij} .$$

In a similar way we also define from φ_{ij} an element

$$\psi_{ji} = t_i\, \varphi_{ij} \in A^*(B_j) \otimes A^*(g^{-1} B_j)$$

with

$$d_0\, \psi_{ji} = t_i\, d_0 \varphi_{ij} .$$

If v_i, v_j are vertices not spanning any 1-simplex of B, then we simply put

$$\psi_{ij} = 0 \in A^*(B_i) \otimes A^*(g^{-1} B_i) ,$$

and

$$\psi_{ji} = 0 \in A^*(B_j) \otimes A^*(g^{-1} B_j) .$$

Now, we replace each ζ_i by

$$\zeta'_i = \zeta_i - \Sigma\, d_0\, \psi_{ik} \in A^*(B_i) \otimes A^*(g^{-1} B_i) ,$$

the summation being extended over all vertices v_k of B of which only a finite number of terms is non-zero.

For each pair of vertices v_i, v_j spanning a 1-simplex of B we have now

$$\begin{aligned} & \mathrm{Rest}_{ij}\, \zeta'_i - \mathrm{Rest}_{ij}\, \zeta'_j \\ &= \mathrm{Rest}_{ij}\, \zeta_i - \mathrm{Rest}_{ij}\, \zeta_j - \Sigma\, \mathrm{Rest}_{ij}\, (t_k d_0 \varphi_{ik} - t_k d_0 \varphi_{jk}) \\ &= \mathrm{Rest}_{ij}\, \zeta_i - \mathrm{Rest}_{ij}\, \zeta_j - \Sigma\, t_k\, (\mathrm{Rest}_{ij}\, \zeta_i - \mathrm{Rest}_{ij}\, \zeta_j) \\ &= 0 , \end{aligned}$$

since $\Sigma\, t_k = 1$ on B_{ij}.

It follows that the collection of ζ'_i for all vertices $v_i \in B$ defines an element ζ' of E_0. Moreover, $d_0\, \zeta' = d_0\, \zeta = 0$ so that $\zeta' \in Z(E_0)$, and passing to homology it becomes $\tilde{\eta}$. Hence, π is onto. In a similar way we prove that any element in the kernel of π is in the image of i, the proof of which will be omitted.

This proves Lemma 3 and also completes the proof of Lemma 2 .

Now, the two spectral sequences arising from filtrations of $A^*(Y)$ and $\bar{A}^*(Y)$ are in fact isomorphis to each other. To see this, let us define a DGA-morphism

$$(7) \qquad h : \bar{A}^*(Y) \to A^*(Y)$$

as follows.

Let $\eta \in \bar{A}^{r,s}(Y) \subset \bar{A}^*(Y)$. To any $\tilde{\sigma} \in Y$, let $g\tilde{\sigma} = \sigma \in B$ with $i_{\tilde{\sigma}}$ as the inclusion $\tilde{\sigma} \subset g^{-1}\sigma$ and $g_{\tilde{\sigma}}$ as the projection $\tilde{\sigma} \to \sigma$. If for some vertex v_i of B we have $\sigma \in B_i$ and

$$\eta(v_i) = \Sigma\, \alpha_j \otimes \beta_j \in A^r(B_i) \otimes A^s(g^{-1}B_i) ,$$

we set

$$(h\eta)(\tilde{\sigma}) = \Sigma\, g^A_{\tilde{\sigma}} \alpha_j(\sigma) \cdot \beta_j(\tilde{\sigma}) \in A^{r+s}(\tilde{\sigma}) .$$

Clearly, $(h\eta)(\tilde{\sigma})$, $\tilde{\sigma} \in Y$ is chosen independent of v_i , satisfies the compatibility condition and therewith determines a welldefined element $h\eta \in A^{r+s}(Y)$. The morphism $\eta \to h\eta$ clearly is also a DGA-morphism.

<u>Lemma 4</u>: The DGA-morphism h of (7) is an H-isomorphism so that we have

$$I^*(Y) \approx \min \bar{A}^*(Y) .$$

<u>Proof</u>: It is easy to see that h respects filtrations of (1) and (4) , viz.

$$h\, F^p \bar{A}^*(Y) \subset F^p A^*(Y) .$$

It follows that h induces morphisms of corresponding Leray spectral sequences

$$h_r : E^p_r \bar{A}^*(Y) \to E^p_r A^*(Y) .$$

From Lemmas 1 and 2 we have

$$h_2 : E^p_2 \bar{A}^*(Y) \approx E^p_2 A^*(Y) .$$

From the isomorphism theorem of spectral sequences we get h_H as a module-isomorphism. Since h preserves multiplications, the isomorphism h_H is also multiplicative or h is an H-isomorphism. It follows that

$$I^*(Y) \approx \min A^*(Y) \approx \min \bar{A}^*(Y) ,$$

as to be proved.

VII.3 FIBER-SQUARE-CONSTRUCTIONS AND FIBER-SQUARE-THEOREM FOR I^*-MEASURES

Definition: Give a fibration

$$(\mathcal{F}) \qquad F \subset Y \xrightarrow[g]{} B$$

and a continuous map

$$f : X \to B$$

which induces a fibration

$$(f^*\mathcal{F}) \qquad F \subset Z \xrightarrow[\bar{g}]{} X$$

with fiber space Z. Then, the commutative diagram of maps

$$(\mathcal{S}) \qquad \begin{array}{ccc} F \subset Z & \xrightarrow{\bar{f}} & Y \\ \bar{g}\downarrow & & \downarrow g \\ X & \xrightarrow{f} & B \end{array}$$

will be said to form a fiber-square, and Z will be said to be determined by the fiber-square-construction from fibration $(\mathcal{F})$ and map f. If $(\mathcal{F})$ and f both are simplicial, so that $(f^*\mathcal{F})$ also is simplicial, then the fiber-square will also be said to be simplicial.

Notation: The fiber-square-constructed space Z is sometimes also denoted by

$$Z = X \times_B Y$$

meaning the subspace of the topological product $X \times Y$, consisting of points (x, y) with $x \in X$, $y \in Y$ and $f(x) = g(y)$.

For a fiber square the homology of the fiber-square-constructed space Z is related to those of X, B, Y and interrelated morphisms by a spectral sequence due to Eilenberg-Moore. In more details, for any coefficient field $\underline{k}$, there is a spectral sequence

$$E_2 \approx \mathrm{Tor}(f^H, g^H) \Longrightarrow H^*_{\underline{k}}(Z).$$

In case the spectral sequence collapses, so that $E_2 \approx H^*_{\underline{k}}(Z)$, then it serves to determine $H^*_{\underline{k}}(Z)$ from the knowledge of diagram

$$(D)_H \qquad H^*_{\underline{k}}(X) \xleftarrow{f^H} H^*_{\underline{k}}(B) \xrightarrow{g^H} H^*_{\underline{k}}(Y).$$

In other words, $H^*_{\underline{k}}$ will be calculable with respect to the fiber-square-construction in such collapse cases.

Cases for which the Eilenberg-Moore spectral sequence collapses, have been considered by Baum, Smith, et al. However, in general this is not true, and H^*-measure of Z cannot be determined in this manner, except for very particular cases. In other words, H^*_k-measure is <u>not calculable</u> with respect to the fiber-square-construction. On the other hand, as we shall show in this and in the next sections, I^*-measure is calculable with respect to the fiber-square-construction, or $I^*(Z)$ is completely determined from the knowledge of diagram

$$(D)_I \qquad I^*(X) \xleftarrow{f^I} I^*(B) \xrightarrow{g^I} I^*(Y).$$

Passing to homology, this implies in particular that, in the case of $\underline{k}$ being of characteristic 0, $H^*(Z) = H^*_{\underline{k}}(Z)$ will be completely determined by $(D)_I$, but not at all by $(D)_H$ alone. Moreover, explicit formulae for the determination of $I^*(Z)$ and in particular $H^*(Z)$ in terms of $(D)_I$ can be given, which is impossible by means of $(D)_H$ alone, except for very special cases.

To state the final result, let us remark that f^I and g^I in $(D)_I$ will give rise to some right resp. left operations to make $I^*(X)$ resp. $I^*(Y)$ a right- resp. a left-DGA over $I^*(B)$, defined as follows:

$$R(f^I): \begin{cases} I^*(X) \otimes I^*(B) \to I^*(X), \\ \xi \otimes \beta \to \xi \cdot f^I\beta \quad (\text{or simply } \xi\beta). \end{cases}$$

$$L(g^I): \begin{cases} I^*(B) \otimes I^*(Y) \to I^*(Y), \\ \beta \otimes \eta \to g^I\beta \cdot \eta \quad (\text{or simply } \beta\eta). \end{cases}$$

With $I^*(X)$ thus considered a right-DGA on $I^*(B)$, we can form – as in § VII.1 with $i: B \to B$ the identity – a DGA $\mathrm{Bar}(f^I, i^I) \approx \mathrm{Bar}^{R(f^I)}(I^*(X))$ which also is a right-DGA over $I^*(B)$. As in § VII.1, form again the DGA

$$\mathrm{Bar}(f^I, g^I) \approx \mathrm{Bar}(f^I, i^I) \underset{I^*(B)}{\otimes} I^*(Y) \approx \mathrm{Bar}(f^I, i^I) \otimes I^*(Y)/I\,,$$

in which I is the sub-DGA generated by elements of the form

$$\zeta\beta \otimes \eta - \zeta \otimes \beta\eta$$

with $\zeta \in \mathrm{Bar}(f^I, i^I)$, $\beta \in I^*(B)$ and $\eta \in I^*(Y)$.

The final result may then be stated in the following

Theorem 1: The I^*-measure is calculable with respect to the fiber-square-constructions for simplicial fiber squares. More precisely, for a simplicial fiber-square $(\mathcal{S})$ $I^*(Z)$ is completely determined by the knowledge of $(D)_I$ and is explicitely given by

(1) $$I^*(Z) \approx \min \mathrm{Bar}(f^I, g^I)\,.$$

We shall postpone the proof of the theorem to the next section, and - for the moment - we shall restrict ourselves to the derivation of some consequences of this theorem. As an immediate corollary - by passing to homology - we get from (1) the following

Theorem 2: For a simplicial fiber square $(\mathcal{S})$ the H^*-measure of the fiber-square-constructed space Z is completely determined by the I^*-measures of the various spaces as well as the interrelated DGA-morphisms, as shown in $(D)_I$. More precisely, we have

(2) $$H^*(Z) \approx \mathrm{Tor}(f^I, g^I)\,,$$

or in current notations,

(2)' $$H^*(Z) \approx \mathrm{Tor}_{I^*(B)}(I^*(X), I^*(Y))\,.$$

Since the ordinary fibration

$(\mathcal{F})$ $$F \subset Y \xrightarrow{\;g\;} B$$

can also be considered a fiber-square (i_0 = inclusion)

$$\begin{array}{ccc} F \subset F & \xrightarrow{\;i_0\;} & Y \\ {\scriptstyle \bar{g}}\big\downarrow & & \big\downarrow{\scriptstyle g} \\ \mathrm{pt.} & \xrightarrow{\;i_0\;} & B \end{array}$$

as a corollary, we get the following

Theorem 3: The I^*-measure and in particular the H^*-measure of the fiber in a simplicial fibration is completely determined by the knowledge of the DGA-morphism

$$g^I : I^*(B) \to I^*(Y)$$

by means of the following formulae:

$$(3) \qquad I^*(F) \approx \min \mathrm{Bar}\,(i_0^I, g^I),$$

$$(4) \qquad H^*(F) \approx \mathrm{Tor}\,(i_0^I, g^I),$$

or in current notation,

$$(4)' \qquad H^*(F) \approx \mathrm{Tor}_{I^*(B)}(\underline{k}, I^*(Y)).$$

Once again, we remark that in all the above theorems for the complete determination of $I^*(Z)$ or $I^*(F)$, even for $H^*(Z)$ or $H^*(F)$, I^*-measures, but not the H^*-measures of B, X and Y should be used. Only in certain cases I^*-measures of B, X or Y coincide with the H^*-measures; then we can replace I^*-measures by H^*-measures for the determination of those of Z or F.

Such cases occur when the space involved is e.g.

(a) A Lie group.

(b) A compact symmetric space.

(c) A homogeneous space $M = G/H$ with G a compact connected Lie group and H a closed connected sub-group of maximal rank or of deficiency 0.

(d) The classifying space of a compact Lie group.

(e) Formal spaces like kompact Kähler manifolds, etc.

We shall take advantage of these facts and deduce some consequences, as follows.

Let G be a compact connected Lie group and H a closed connected subgroup of G. For fibrations with the homogeneous space $M = G/H$ as fiber and G as structural group we have the universal one:

$$(\mathcal{U}) \qquad G/H \subset B_H \xrightarrow{g} B_G,$$

in which B_H, B_G are classifying spaces with I^*- or H^*-measures polynomial

DGA's. Let

(5) $\qquad I^*(B_G) \approx \mathrm{Polym}(s_1, \ldots, s_N)$,

(6) $\qquad I^*(B_H) \approx \mathrm{Polym}(s_{H_1}, \ldots, s_{HL})$,

(7) $\qquad I^*(G) \approx \mathrm{Extr}(p_1, \ldots, p_N)$,

in which p_μ are primitive elements with s_μ as transgressions. Rewrite the above fibration in the form of a fiber-square

$$\begin{array}{ccc} G/H & \longrightarrow & B_H \\ \downarrow & & \downarrow g \\ \mathrm{pt.} & \xrightarrow{i_0} & B_G \end{array}$$

Then, Theorem 3 gives

(8) $\qquad I^*(G/H) \approx \min(\mathrm{Bar}(i_0^I, g^I))$.

By Theorem 2 of § VII.2 there is a DGA-morphism of $\mathrm{Kosz}(i^I, g^I)$ in $\mathrm{Bar}(i^I, g^I)$ which induces an H-isomorphism. Hence, we also get

(8)' $\qquad I^*(G/H) \approx \min(\mathrm{Kosz}(i_0^I, g^I))$.

In the present case we see that

(9) $\qquad \mathrm{Kosz}(i_0^I, g^I) \approx \mathrm{Extr}(p_1, \ldots, p_N) \otimes \mathrm{Polym}(s_{H1}, \ldots, s_{HL})$

with a differential given by

(10) $$\begin{cases} dp_\mu = g^I s_\mu , \\ ds_{H\lambda} = 0 . \end{cases}$$

We see that this is just the Cartan algebra associated to the homogeneous space G/H , as defined in Chap. V . Hence, we have deduced the Cartan theorem from the fiber-square theorem with an alternative formulation, in the following

Theorem 4: Let G be a compact connected Lie group and H a closed connected subgroup, then the I^*-measure of the homogeneous space G/H is given by

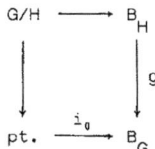

$$(11) \qquad I^*(G/H) \approx \min(\mathrm{Kosz}(i_0^I, g^I))$$
$$\approx \min(\mathrm{Kosz}(i_0^H, g^H)),$$

where $i_0 : pt. \subset B_G$ is the inclusion and $g : B_H \to B_G$ is the projection of the universal fibration $(\mathcal{U})$. In passing to homology we have in particular

$$(11)' \qquad H^*(G/H) \approx \mathrm{Tor}(i_0^I, g^I)$$
$$\approx \mathrm{Tor}(i_0^H, g^H).$$

Consider now a general fibration with fiber G/H and a structural group G, as before:

$$(\mathcal{E}) \qquad G/H \subset E \to B.$$

Let the fibration be induced by the universal one, by a map

$$f : B \to B_G$$

so that we have a fiber square:

$$\begin{array}{ccc} G/H \subset E & \longrightarrow & B_H \\ \downarrow & & \downarrow g \\ B & \xrightarrow{f} & B_G \end{array}$$

By the fiber square theorem we then can determine $I^*(E)$ and $H^*(E)$ by means of the diagram

$$I^*(B) \xleftarrow{f^I} I^*(B_G) \xrightarrow{g^I} I^*(B_H),$$

as before.

Now, for the classifying spaces B_H and B_G the I^*-measure coincides with the H^*-measure. If B is certain space for which the I^*-measure coincides also with the H^*-measure, then – in the above theorem – we can replace all the I^*-measures of B_H, B_G and B by H^*-measures. Thus, we get the following results due to Baum, Smith, Wolf, et al.

Theorem 6: (Baum–Smith) If the base space of the fibration $\mathcal{E}$ is a Riemannian symmetric space then

$$(12) \qquad I^*(E) \approx \min \mathrm{Bar}(f^H, g^H),$$

and

(13) $$H^*(E) \approx \mathrm{Tor}\,(f^H, g^H)$$

or in current notations,

(13)' $$H^*(E) \approx \mathrm{Tor}_{H^*(B_G)}\,(H^*(B), H^*(B_H))\ .$$

Theorem 6: (Wolf) If the base space of the fibration ($\mathcal{E}$) itself is a homogeneous space G'/H' of which G' is a compact connected Lie group, and H' is a closed connected subgroup of deficiency 0 or of maximal rank, then again we have (12) and (13) or (13)' .

VII.4 PROOF OF THE FIBER-SQUARE-THEOREM

Consider the fiber-square

($\mathcal{S}$) $$\begin{array}{ccc} F \subset Z & \xrightarrow{\bar{f}} & Y \\ \bar{g}\downarrow & & \downarrow g \\ X & \xrightarrow{f} & B \end{array}$$

as in § VII.3 . We have then induced DGA-morphisms

$(D)_A$ $$A^*(X) \xleftarrow{f^A} A^*(B) \xrightarrow{g^A} A^*(Y)$$

and

$(D)_I$ $$I^*(X) \xleftarrow{f^I} I^*(B) \xrightarrow{g^I} I^*(Y)$$

The idea of the proof is to get first some expression of $I^*(Z)$ in terms of diagram $(D)_A$, and then deduce the expression in terms of $(D)_I$, which is the required one.

Let $V = \{v_i\}$ and $W = \{w_j\}$ be the sets of vertices of B and X respectively. In § VII.3 we have defined, for the fibration

($\mathcal{F}$) $$F \subset Y \xrightarrow{g} B$$

and, hence, also for the induced fibration

($f^*\mathcal{F}$) $$F \subset Z \xrightarrow{\bar{g}} X$$

some DGA's

$$\bar{A}^*(Y) = \Sigma \bar{A}^q(Y), \quad \bar{A}^q(Y) = \sum_{r+s=q} \bar{A}^{r,s}(Y),$$

$$\bar{A}^*(Z) = \Sigma \bar{A}^q(Z), \quad \bar{A}^q(Z) = \sum_{r+s=q} \bar{A}^{r,s}(Z),$$

$$\bar{A}^{r,s}(Y) = \{\eta(v_i) \in A^r(B_i) \otimes A^s(g^{-1}B_i), \quad v_i \in V/(C_Y)\},$$

$$\bar{A}^{r,s}(Z) = \{\zeta(w_j) \in A^r(X_j) \otimes A^s(\bar{g}^{-1}X_j), \quad w_j \in W/(C_Z)\},$$

in which $(C_Y), (C_Z)$ denote some natural compatibility conditions and X_j the subcomplex of X analogous to B_i of B, described in § VII.3 .

Now, define a DGA-morphism

$$\iota_B : A^*(B) \to \bar{A}^*(Y)$$

simply by setting for any $\beta \in A^r(B)$ and any vertex v_i of B,

$$(\iota_B\beta)(v_i) = g^A\beta(B_i) \otimes 1 \in A^r(B_i) \otimes A^0(g^{-1}B_i).$$

By § VII.1 , ι_B, f^A will then turn $\bar{A}^*(Y)$ into a left-DGA and $A^*(X)$ into a right-DGA over $A^*(B)$ respectively, so that we can define a quotient DGA

$$A^*(X) \underset{A^*(B)}{\otimes} \bar{A}^*(Y)$$

from $A^*(X) \otimes \bar{A}^*(Y)$ with projection denoted by π, say.

Then, define DGA-morphisms

$$\gamma : \bar{A}^*(Y) \to \bar{A}^*(Z),$$

$$\lambda : A^*(X) \otimes \bar{A}^*(Y) \to \bar{A}^*(Z)$$

as follows.

Let $w_j \in W$, $f(w_j) = v_i \in V$. Let $\eta \in \bar{A}^*(Y)$ and $\eta(v_i) = \Sigma\, \alpha_{is} \otimes \beta_{is} \in$ $\in A^*(B_i) \otimes A^*(g^{-1}B_i)$. Set by definition

$$(\gamma(\eta))(w_j) = f_j^A \alpha_{is} \otimes \bar{f}_j^A \beta_{is} \in A^*(X_j) \otimes A^*(\bar{g}^{-1}X_j),$$

in which

$$f_j : X_j \to B_i, \quad \bar{f}_j : \bar{g}^{-1}X_j \to g^{-1}B_i$$

are simplicial maps which are restrictions of f and $\bar{f}$ to X_j and $\bar{g}^{-1}X_j$ respectively. It is clear that $\gamma(\eta)$ is a well-defined element of $\bar{A}^*(Z)$, and γ is a DGA-morphism. For the definition of λ we simply set then

$$\lambda(\xi \otimes \eta) = \iota_X(\xi) \cdot \gamma(\eta)$$

for any $\xi \in A^*(X)$, $\eta \in \bar{A}^*(Y)$, where

$$\iota_X : A^*(X) \to \bar{A}^*(Z)$$

is defined as ι_B .

With $A^*(X)$ as right DGA over $A^*(B)$ by f^A , we can construct by § VII.1 $\mathrm{Bar}^{R(f^A)}(A^*(X))$ as right-DGA over $A^*(B)$ and form the quotient algebra

$$P^* = \mathrm{Bar}^{R(f^A)}(A^*(X)) \underset{A^*(B)}{\otimes} \bar{A}^*(Y) .$$

There is also a natural DGA-morphism

$$\varepsilon : \mathrm{Bar}^{R(f^A)}(A^*(X)) \to A^*(X) ,$$

defined by $\varepsilon(\xi[\,]\beta) = \xi \cdot f^A \beta$, while $\varepsilon(\xi[\beta_1 | \ldots | \beta_n]\beta) = 0$

for $n \geq 1$, in which $\xi \in A^*(X)$, $\beta_i \in (A^*(B))^+$, and $\beta \in A^*(B)$.

From λ , π and ε we can then complete the following commutative diagram by introducing natural DGA-morphisms μ and θ :

$$\begin{array}{ccc}
P^* = \mathrm{Bar}^{R(f^A)}(A^*(X)) \underset{A^*(B)}{\otimes} \bar{A}^*(Y) & \overset{\theta}{\searrow} & \\
\downarrow \varepsilon \otimes 1 & & \\
A^*(X) \underset{A^*(B)}{\otimes} \bar{A}^*(Y) & \xrightarrow{\ \mu\ } & \bar{A}^*(Z) \\
\uparrow \pi & & \\
A^*(X) \otimes \bar{A}^*(Y) & \overset{\lambda}{\nearrow} &
\end{array}$$

By § VII.3 , associating to the fibrations $(\mathcal{F})$ and $(f^*\mathcal{F})$ we have filtrations

$$F^p \bar{A}^*(Y) = \sum_{r \geq p} \bar{A}^{r,*}(Y) ,$$

$$F^p \bar{A}^*(Z) = \sum_{r \geq p} \bar{A}^{r,*}(Z)$$

with

$$E_2^p \bar{A}^*(Y) \approx H^*(B) \otimes H^*(F) \Longrightarrow H^*(Y) ,$$

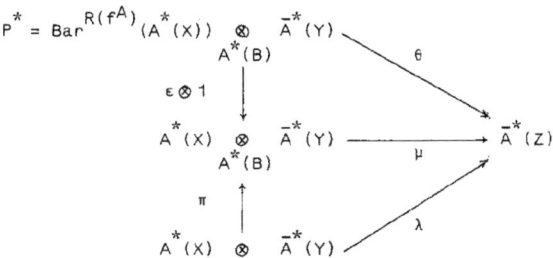

and

$$E_2^p \bar{A}^*(Z) \approx H^p(X) \otimes H^*(F) \Longrightarrow H^*(Z) .$$

Now, filtrate P^* by

$$F^p P^* = \sum_i [\mathrm{Bar}^{R(f^A)}(A^*(X))]^i \underset{A^*(B)}{\otimes} F^{p-i}\bar{A}^*(Y) .$$

Then we have successively:

$$E_0^p P^* \approx \sum_i [\mathrm{Bar}^{R(f^A)}(A^*(X))]^i \underset{A^*(B)}{\otimes} E_0^{p-i}\bar{A}^*(Y) , \qquad d_0 = d_0 \mid E_0\bar{A}^*(Y) ,$$

$$E_1^p P^* \approx \sum_i [\mathrm{Bar}^{R(f^A)}(A^*(X))]^i \underset{A^*(B)}{\otimes} A^{p-i}(B) \otimes H^*(F)$$

$$\approx [\mathrm{Bar}^{R(f^A)}(A^*(X))]^p \otimes H^*(F) , \qquad d_1 = d \mid \mathrm{Bar}^{R(f^A)}(A^*(X)) ,$$

$$E_2^p P^* \approx H^p(X) \otimes H^*(F) ,$$

and

$$E_2^p P^* \Longrightarrow H(P^*) .$$

For the DGA-morphism $\theta : P^* \to \bar{A}^*(Z)$ we have

$$\theta F^p P^* \subset F^p \bar{A}^*(Z)$$

so that θ induces morphisms of spectral sequences

$$\theta_r : E_r P^* \to E_r \bar{A}^*(Z) .$$

Since

$$\theta_2 : E_2 P^* \approx E_2 \bar{A}^*(Z) \; [\approx H^*(X) \otimes H^*(F)]$$

we see that additively

$$\theta_H : H(P^*) \approx H(\bar{A}^*(Z)) \; [\approx H^{\oplus}(Z)] .$$

Since θ is multiplicative, θ_H is also a multiplicative isomorphism or θ is an H-isomorphism. It follows that

$$I^*(Z) \approx \min \bar{A}^*(Z) \approx \min P^*$$

or

$$I^*(Z) \approx \min (\mathrm{Bar}^{R(f^A)}(A^*(X)) \underset{A^*(B)}{\otimes} \bar{A}^*(Y)) .$$

In particular, we have in passing to homology,

$$H^*(Z) \approx \mathrm{Tor}\,(f^A, \iota_B) \approx \mathrm{Tor}_{A^*(B)}\,(A^*(X), \bar{A}^*(Y))\,.$$

The last two formulae give explicit expressions for the determination of the I^*-measure and, hence, also of the H^*-measure of Z from diagram $(D)_A$, in terms of the A^*-measures of various spaces involved. However, as we have pointed out in § I.4, A^*-measures are not desirable, so we have to replace the A^*-measures in these formulae by measures meeting the requirements stated in § I.4, here viz. the I^*-measures of the various spaces.

For this purpose, let us remark that $I^*(Y) \approx \min \bar{A}^*(Y)$ by § VII.2, and thus consider the following commutative diagram of DGA-morphisms:

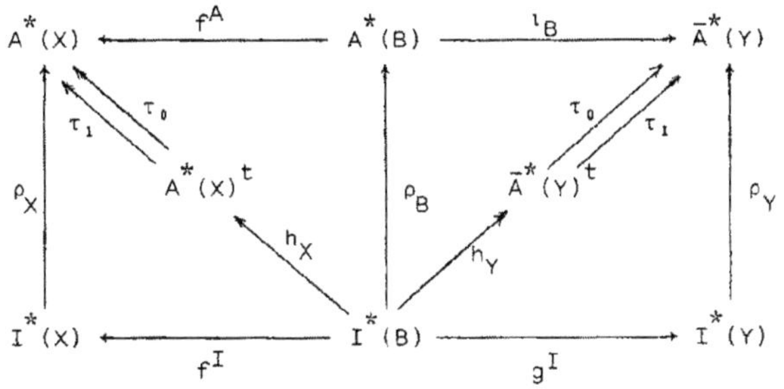

In the diagram ρ_X, ρ_B, ρ_Y are associated minimal morphisms, and h_X, h_Y, etc. arise from the $\approx$-commutativity of the DGA-morphisms $\rho_X f^I \approx f^A \rho_B$ and $\rho_Y g^I \approx \iota_B \rho_B$.

From the above commutative diagram of DGA-morphisms we get then the following commutative one

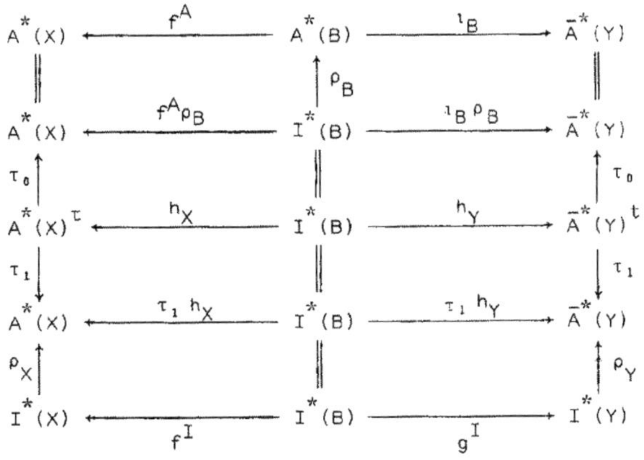

Since each vertical arrow is either an identity or an H-isomorphism, we have by § VII.1 a sequence of isomorphisms

$$
\begin{aligned}
&\min\left(\mathrm{Bar}^{R(f^A)}(A^*(X)) \underset{A^*(B)}{\otimes} \bar{A}^*(Y)\right) \\
\approx\ & \cdots \\
\approx\ & \min\left(\mathrm{Bar}^{R(f^I)}(I^*(X)) \underset{I^*(B)}{\otimes} I^*(Y)\right)
\end{aligned}
$$

in which the expressions omitted are analog to those obtained from the middle rows of the last diagram. From this we immediately get the theorem to be proved.

VII.5 FIBER-SPACE THEOREM AND OTHER APPLICATIONS

Consider a simplicial fibration

$$(\mathcal{F}) \qquad F \subset Y \xrightarrow{\ g\ } B$$

as a fiber square

$$
\begin{array}{ccc}
F \subset F & \xrightarrow{\ \bar{f}\ } & Y \\
\bar{g}\downarrow & & \downarrow g \\
\mathrm{pt.} & \xrightarrow{\ f\ } & B
\end{array}
$$

in which $f, \bar{f}$ are inclusions. By the fiber-square theorem we have then

$$(1) \qquad I^*(F) \approx \min A^*(F)\ ,$$

where

$$(2) \qquad A^*(F) \approx \mathrm{Bar}\,(f^I, g^I) \approx \Omega(I^*(B)) \otimes I^*(Y)\ .$$

The differential on the right-hand side, or the one in $A^*(F)$, to be written as d_F for the sake of distinguishing with the others, is in the form $(b_j \in I^*(B)$, $\deg b_j > 0$, $y \in A^*(F))$:

$$
(3) \qquad
\begin{aligned}
d_F([b_1|\ldots|b_n]y) = & \ \Sigma\,(-1)^{d_i}[b_1|\ldots|b_i\, b_{i+1}|\ldots|b_n]\,y \\
& + (-1)^{d_n}\cdot[b_1|\ldots|b_{n-1}]\,b_n\, y \\
& - \Sigma\,(-1)^{d_{i-1}}\cdot[b_1|\ldots|db_i|\ldots|b_n]\,y \\
& + (-1)^{d_n}\cdot[b_1|\ldots|b_n]\,dy\ ,
\end{aligned}
$$

in which

(4) $$d_i = \sum_{j=1}^{i} (\deg b_j - 1), \qquad i = 1, \ldots, n$$

and

$$b_n y = g^I b_n \cdot y .$$

Let $i_B : B \to B$ be the identity map inducing the identity morphism

$$i_B^I : I^*(B) \to I^*(B) .$$

Then,

(5) $$\begin{aligned} \mathrm{Bar}(i_B^I, g^I) &\approx I^*(B) \otimes \mathrm{Bar}(f^I, g^I) \\ &\approx I^*(B) \otimes A^*(F) \end{aligned}$$

will have a differential d in the form

(6) $$d = d_\otimes + d_\tau ,$$

where $d_\otimes$ is just the differential in $I^*(B) \otimes A^*(F)$ as a tensor product, while d_τ is given by

(7) $$d_\tau(b\,[b_1 | \ldots | b_n]\,y) = (-1)^{\deg b} \cdot bb_1[b_2 | \ldots | b_n]\,y .$$

Such a differential d, as in § II.2, is said to be a <u>twisted</u> one in the tensor product $I^*(B) \otimes A^*(F)$ with the <u>twisted part</u> d_τ. Thus, we write the tensor product in the form $I^*(B) \underset{\tau}{\otimes} A^*(F)$ to make this fact evident. Applying the fiber-square theorem to the trivial fiber-square

$$\begin{array}{ccc} F \subset Y & \longrightarrow & Y \\ \downarrow & & \downarrow g \\ B & \xrightarrow{i_B} & B \end{array}$$

we get then

$$I^*(Y) \approx \min \mathrm{Bar}(i_B^I, g^I)$$

or

(8) $$I^*(Y) \approx \min (I^*(B) \underset{\tau}{\otimes} A^*(F)) .$$

The above formula shows in a quite concise form that the I^*-measure of the fiber-space Y can be determined by the I^*-measure of B and the A^*-measure of F with a certain twisted differential in the corresponding tensor product. However, since the A^*-measure is not a desirable one, we have to replace the A^*-measure by some more adequate ones, viz. the I^*-measure of F. In fact, we have the following theorem which already has been proved in § VI.5. Here, we have an alternative proof as a consequence of the above fiber-square-theorem, by means of theorems of Moore in § II.6. Now, we restate the theorem in question, as follows.

Theorem 1: (Fiber-Space Theorem) For a simplicial fibration

$$F \subset Y \xrightarrow{g} B$$

the I^*-measure of the fiber space Y is given by

$$(9) \qquad I^*(Y) \approx \min (I^*(B) \otimes_\tau I^*(F)),$$

in which $I^*(B) \otimes_\tau I^*(F)$ is the tensor procuct with a certain twisted differential

$$(10) \qquad d = d_\otimes + d_\tau .$$

In particular, we have

$$(11) \qquad H^*(Y) \approx H(I^*(B) \otimes_\tau I^*(F)).$$

The above Fiber-Space Theorem shows how the I^*- and, hence, also H^*-measure of a fiber space can be determined by the I^*-measures of the base and the fiber, as far as some twisted differential is known. In § VI.6 we have shown that, in some important cases, such twisted differentials can be determined explicitly. Further examples will be given below. We start with the following

Definition: Consider an universal fibration

$$(\mathcal{U})_F \qquad F \subset E_F \to B_G$$

with structural group G, classifying space B_G, and fiber F on which G operates. Let

$$(\mathcal{F})_F \qquad F \subset E \to B$$

be the fibration induced by a map

$$f : B \to B_G$$

from the universal fibration $(\mathcal{U})_F$. Then, for any element $x \in I^*(B_G)$, the element $f^I x \in I^*(B)$ where $f^I : I^*(B_G) \to I^*(B)$ is a DGA-morphism induced by f will be called the characteristic element corresponding to the element x and the induced morphism f^I.

If $dx = 0$ then passing to homology the class of $f^I x$ in $H(B)$ is uniquely determined by the class of x in $H(B_G)$ and is the usual characteristic class of the fibration $(\mathcal{F})_F$ corresponding to the homology class of x. In particular, for a fibration with orthogonal groups as structural group, we may speak of Euler element and Pontrjagin elements corresponding to the usual Euler class and Pontrjagin classes of the fibration and, similarly, for Chern elements for fibrations with unitary groups as structural group.

Let G/H be a homogeneous space with G a compact connected Lie group and H a closed connected subgroup. Let

$$(\mathcal{F}) \qquad G/H \subset E \to B$$

be a fibration with fiber G/H and structural group G, induced by

$$f : B \to B_G$$

from the universal fibration

$$(\mathcal{U}) \qquad G/H \subset B_H \to B_G .$$

As in § V.5, we have

$$I^*(G) \approx \mathrm{Extr}\,(p_1, \ldots, p_N)$$

with p_μ primitive elements and

$$I^*(B_G) \approx \mathrm{Polym}\,(s_1, \ldots, s_N)$$

with s_μ some definite transgression of p_μ. Since $I^*(B_G)$ is a polynomial algebra, we get from the Fiber-Square Theorem by using theorems of § VII.1

$$I^*(G/H) \approx \min \mathrm{Kosz}\,(i_0^I, g^I) ,$$

$$I^*(E) \approx \min \mathrm{Kosz}\,(f^I, g^I) .$$

By § VII.1, we see that

$$\mathrm{Kosz}\,(i_0^I, g^I) \approx \mathrm{Extr}\,(x_1, \ldots, x_N) \otimes I^*(B_H)$$

with a differential

$$dx_\mu = -g^I s_\mu, \quad d|I^*(B_H) = 0.$$

It follows that, by taking primitives p_μ of $I^*(G) \approx H^*(G)$ with s_μ as transgressives and setting

$$x_\mu = -p_\mu,$$

the DGA $\mathrm{Kosz}\,(i_0^I, g^I)$ is seen to be just the Cartan algebra C associated to the pair (G,H), as described in § V.4.

Similarly, we have

$$\mathrm{Kosz}\,(f^I, g^I) \approx I^*(B) \otimes \mathrm{Extr}\,(x_1, \ldots, x_N) \otimes I^*(B_H)$$

with a differential d given by

$$dx_\mu = f^I s_\mu - g^I s_\mu,$$

and with d on $I^*(B)$ and $I^*(B_H)$ the original ones. Comparing with the Cartan algebra and the corresponding differential we see, that

$$\mathrm{Kosz}\,(f^I, g^I) \approx I^*(B) \underset{\tau}{\otimes} C$$

with differential $d = d_\otimes + d_\tau$ of which the twisted part d_τ is given by

$$d_\tau p_\mu = -f^I s_\mu,$$

$$d_\tau \mid I^*(B_H) = 0.$$

The twisted part d_τ is thus determined by $f^I s_\mu$, the characteristic elements of the fibration (F) corresponding to s_μ and f^I.

Let

$$\rho : I^*(G/H) \to C$$

be the minimal morphism, then we have

$$I^*(E) \approx \min \mathrm{Kosz}\,(f^I, g^I)$$
$$\approx \min\,(I^*(B) \otimes_\tau C)$$
$$\approx \min\,(I^*(B) \otimes_\tau I^*(G/H))$$

for which the twisted part in the last tensor product is the one induced by that of $I^*(B) \otimes_\tau C$ under ρ, or

$$d_\tau\, x = d_\tau\, \rho x$$

for any $x \in I^*(G/H)$.

The result can thus be formulated in the following form:

<u>Theorem 2</u>: For a fibration $(\mathcal{F})$ with fiber a homogeneous space G/H as above, the I^*-measure, and hence also the H^*-measure, of the fiber space is completely determined by the I^*-measure of the base, the Cartan algebra of the homogeneous space and the characteristic elements $f^I s_\mu$ of the fibration. In more details

$$I^*(E) \approx \min\,(I^*(B) \otimes_\tau I^*(G/H)),$$
$$H^*(E) \approx H\,(I^*(B) \otimes_\tau I^*(G/H)),$$

with a twisted differential in the tensor product of which the twisted part is determined by the characteristic elements.

Now, consider the particular case of a sphere bundle with fiber an m-sphere as a homogeneous space

$$S^m = SO(m+1)/\,SO(m).$$

We have thus an universal bundle

$$(\mathcal{U})_m \qquad S^m \subset B_{SO(m)} \xrightarrow{\;g\;} B_{SO(m+1)}.$$

First let $m = 2n-1$ be odd. Then

$$I^*(SO(2n)) \approx \mathrm{Extr}\,(p_1, \ldots, p_n),$$
$$I^*(B_{SO(2n)}) \approx \mathrm{Polym}\,(s_1, \ldots, s_n),$$
$$I^*(B_{SO(2n-1)}) \approx \mathrm{Polym}\,(s_{H1}, \ldots, s_{H,n-1}),$$

in which $(\mu = 1, \ldots, n-1;\quad k = 1, \ldots, n)$

$$\deg p_\mu = 4\mu - 1 , \qquad \deg p_n = 2n - 1 ,$$

$$\deg s_\mu = \deg s_{H\mu} = 4 \quad , \qquad \deg s_n = 2n ,$$

$$s_k = \text{transgression of } p_k ,$$

$$g^I s_\mu = s_{H\mu} , \qquad g^I s_n = 0 .$$

The Cartan algebra is given by

$$C \approx \operatorname{Extr}(p_1 ,\dots, p_n) \otimes \operatorname{Polym}(s_{H1} ,\dots, s_{H,\, n-1})$$

with differential d_C given by

$$d_C\, p_\mu = g^I s_\mu = s_{H\mu} , \qquad d_C\, p_n = 0 ,$$

$$d_C\, s_{H\mu} = 0 .$$

The I^*-measure of S^{2n-1} thus is

$$I^*(S^{2n-1}) \approx \min C \approx \operatorname{Free}(x)$$

with a single generator x of degree $2n - 1$, and the minimal morphism ρ is given by

$$\rho x = p_n .$$

Hence, for a sphere-fibration

$$(\mathcal{S})_{2n-1} \qquad S^{2n-1} \subset E \xrightarrow{\;g\;} B$$

induced by a map $f : B \to B_{SO(2n)}$ from $(\mathcal{U})_{2n-1}$ we have

$$I^*(E) \approx \min (I^*(B) \underset{\tau}{\otimes} I^*(S^{2n-1}))$$

$$\approx \min (I^*(B) \underset{\tau}{\otimes} \operatorname{Free}(x)) .$$

The twisted part of the differential is given by

$$d_\tau x = - f^I s_n ,$$

where $f^I s_n$ is the characteristic element whose homology class is the Euler class of the bundle.

Next, consider the case $m = 2n$ even. Then,

$$I^*(SO(2n+1)) \approx \mathrm{Extr}\,(p_1, \ldots, p_n),$$

$$I^*(B_{SO(2n+1)}) \approx \mathrm{Polym}\,(s_1, \ldots, s_n),$$

$$I^*(B_{SO(2n)}) \approx \mathrm{Polym}\,(s_{H1}, \ldots, s_{Hn}),$$

in which $(\mu = 1, \ldots, n-1; \quad k = 1, \ldots, n)$

$$\deg p_k = 4k-1, \qquad \deg s_k = 4k,$$

$$\deg s_{H\mu} = 4\mu, \qquad \deg s_{Hn} = 2n,$$

$$s_k = \text{transgression of } p_k,$$

$$g^I s_\mu = s_\mu, \qquad g^I s_n = c_n s_{Hn}^2.$$

The c_n is certain non-zero rational number, dependent on n which does not interest us. In fact, because of the fact that the basic field is of characteristic 0, the generators may be chosen up to non-zero multiples, so that c_n may be absorbed. Thus, the Cartan algebra is given by

$$C \approx \mathrm{Extr}\,(p_1, \ldots, p_n) \otimes \mathrm{Polym}\,(s_{H1}, \ldots, s_{Hn})$$

with differential d_C given by

$$d_C p_\mu = s_{H\mu}, \qquad d_C p_n = c_n s_{Hn}^2,$$

$$d_C s_{Hk} = 0.$$

Thus, we have

$$I^*(S^{2n}) = \min C \approx \mathrm{Free}\,(x, y)$$

with

$$\deg x = 4n-1, \qquad \deg y = 2n,$$

$$dx = c_n y^2, \qquad dy = 0,$$

and the minimal morphism $\rho : I^*(S^{2n}) \to C$ given by

$$\rho x = p_n, \qquad \rho y = s_{Hn}.$$

For a sphere-fibration

$(S)_{2n}$ $\qquad S^{2n} \subset E \xrightarrow{g} B$

induced by a map $f : B \to B_{SO(2n+1)}$ from $(\mathcal{U})_{2n}$ we therefore have

$$I^*(E) \approx \min (I^*(B) \underset{\tau}{\otimes} I^*(S^{2n}))$$

$$\approx \min (I^*(B) \underset{\tau}{\otimes} \mathrm{Free}(x,y)) .$$

The twisted part of the differential is given by

$$d_\tau x = - f^I s_n ,$$

$$d_\tau y = 0 ,$$

in which $f^I s_n$ is the characteristic element of the fibration whose homology class is the ordinary Pontrjagin class in dimension $4n$.

Summarizing up, we have the following

<u>Theorem 3</u>: For a sphere-fibration $(S)_m$ the I^*-measure, and hence also the H^*-measure, of the fiber space E is completely determined by the I^*-measure of the base B and some characteristic element of the fibration. More explicitly, we have:

For $m = 2n - 1$ odd,

$$I^*(E) \approx \min (I^*(B) \underset{\tau}{\otimes} \mathrm{Free}(x)) ,$$

$$H^*(E) \approx H(I^*(B) \underset{\tau}{\otimes} \mathrm{Free}(x)) ,$$

with

$$\deg x = 2n-1 , \qquad dx = -e ,$$

where e is the Euler element of the fibration.

For $m = 2n$ even,

$$I^*(E) \approx \min (I^*(B) \underset{\tau}{\otimes} \mathrm{Free}(x,y)) ,$$

$$H^*(E) \approx H(I^*(B) \underset{\tau}{\otimes} \mathrm{Free}(x,y)) ,$$

with

$$\deg x = 4n-1 , \qquad \deg y = 2n ,$$

$$dx = c_n y^2 - \pi_n ,$$

$$dy = 0 ,$$

where π_n is the Pontrjagin element of degree $4n$ of the fibration.

As an application let us consider e.g. the Thom space T of a sphere fibration $(\mathcal{S})_m$. In fact, let D be the associated disc-fibration of $(\mathcal{S})_m$. Then, D is contractible to B, so that for the inclusion $j : E \subset D$ the induced DGA-morphism $j^I : I^*(D) \to I^*(E)$ is the same as the DGA-morphism

$$g^I : I^*(B) \to I^*(E) \approx \min(I^*(B) \otimes_\tau I^*(S^m)) .$$

Since the Thom space T is homotopically the space obtained from D by erecting a cone over the subspace E, the method of Chap. VI can be applied to determine $I^*(T)$ from g^I. By the same Chapter, instead of g^I, we can also use the natural inclusion DGA-morphism

$$k : I^*(B) \subset I^*(B) \otimes_\tau I^*(S^m)$$

for the determination of $I^*(T)$.

Thus, consider first the case $m = 2n-1$ odd, so that $I^*(S^m) = \text{Free}(x)$ with $d_\tau x = -e$. As in Chapter VI, form a DGA

$$\tilde{I}^*(B) \approx I^*(B) \otimes \text{Free}(x^0, x^+)$$

with $\deg x^0 = 2n-1$, $\deg x^+ = 2n$ and $dx^0 = x^+$, $dx^+ = 0$. Extend also the DGA-morphism k to one $\tilde{k} : \tilde{I}^*(B) \to I^*(B) \otimes_\tau I^*(S^{2n-1})$ by $\tilde{k} = k | I^*(B)$, $\tilde{k}(x^0) = x$, $\tilde{k}(x^+) = dx = -e$. Then, $(\text{Ker}\,\tilde{k})^+$ is the ideal in $\tilde{I}^*(B)$ generated by $u = x^+ + e$. Since $\min \text{Ker}\,\tilde{k} \approx I^*(T)$ and $H^s(B) \approx H_s(I^*(B)) \approx H_s(\tilde{I}^*(B))$, we see that the module-morphism $z \to zu$ for $z \in I^*(B)$ will give rise to module-isomorphisms $H^s(B) \approx H^{s+2n}(T)$. This is just the Thom isomorphism of the sphere-fibration $(\mathcal{S})_{2n-1}$.

Next, consider the case $m = 2n$ even. Then, for the determination of $I^*(T)$ we have to consider the natural inclusion morphism

$$k : I^*(B) \subset I^*(B) \otimes_\tau \text{Free}(x,y)$$

with x,y given, as in Theorem 3. To facilitate the calculations let us replace $I^*(B) \otimes_\tau \text{Free}(x,y)$, as it is clearly legitimate by § II.6, by another DGA $J^*(B)$ obtained from $I^*(B)$, by adjoining a single generator $\bar{y}$ such that

$$\deg \bar{y} = 2n , \quad c_n \bar{y}^2 = \pi_n , \quad d\bar{y} = 0 .$$

Form now a DGA

$$\tilde{I}^*(B) \approx I^*(B) \otimes \mathrm{Free}\,(y^0, y^+)$$

with $\deg y^0 = 2n$, $\deg y^+ = 2n+1$, $dy^0 = y^+$, $dy^+ = 0$. Extend also the natural inclusion $\bar{k} : I^*(B) \subset J^*(B)$ to a DGA-morphism

$$\tilde{k} : \tilde{I}^*(B) \to J^*(B)$$

such that

$$\tilde{k}(y^0) = \bar{y}, \quad \tilde{k}(y^+) = 0 .$$

It follows that $(\mathrm{Ker}\,\tilde{k})^+$ is an ideal in $\tilde{I}^*(B)$ with generators

$$u = c_n (y^0)^2 - \pi_n, \quad v = y^+$$

such that

$$du = 2c_n y^0 v, \quad dv = 0 .$$

For any element w in $(\mathrm{Ker}\,\tilde{k})^+$ we can write it in the form

$$w = \sum_{i \geq 1} (a_i + a_i' y^0) a^i + \sum_{\geq 0} (b_i + b_i' y^0) u^i v$$

with $b, a', b' \in I^*(B)$ while $a \in I^*(B) \times \mathrm{Free}\,(y^0)$.
Then, w will be a cycle if and only if

$$da_i = 0, \quad da_i' = 0, \quad a_i \sim 0, \quad a_i' \sim C .$$

We have then

$$w \sim zv$$

with z a cycle in $I^*(B) \times \mathrm{Free}\,(y^0, y^+)$. As in the preceding case, it follows that the module-morphism $z \to zv$ will give rise to the Thom-isomorphism $H^s(B) \approx H^{s+2n+1}(T)$ in $(S)_{2n}$.

We may summarize the results as in the following

<u>Theorem 4</u>: For sphere-fibrations the I^*-measure is calculable with respect to the Thom-space construction. More precisely, the I^*-measure, and hence also the H^*-measure, of the Thom-space of a sphere-fibration is completely determined by the I^*-measure (but not the H^*-measure) of the base space as well as by certain

characteristic element of the fibration which, when passing to homology, is either the Euler class or the Pontrjagin class in the top dimension, according to the dimension of the fiber-sphere being odd or even.

BIBLIOGRAPHY

Andrews, P. & Arkowitz, M.

1. Sullivan's minimal models and higher order Whitehead products. Canadian J. Math. 30(1978), 961-982.

Babenko, L. K.

1. Lie algebras of homotopy groups of minimal Sullivan models. Math. Notes 20(1976), 1005-1011.

Baues, H. J. & Lemaire, J. M.

1. Minimal models in homotopy theory. Math. Ann. 225(1977), 219-242.

Baum, P. F.

1. On the cohomology of homogeneous spaces. Topology 7(1968), 15-38.

Baum, P. F. & Smith, L.

1. The real cohomology of differential fiber bundles. Comm. Math. Helv. 42 (1967), 171-179.

Borel, A.

1. Sur la cohomologie des espaces fibrés principaux et des espaces homogènes de groupes de Lie compacts. Annals of Math. 57(1953), 115-207.

Bott, R.

1. Lectures on characteristic classes and foliations. Lectures on algebraic and differential topology. Lecture Notes in Math. No. 279, Springer-Verlag (1972), 1-94.

Bousfield, A. K. & Guggenheim, V. K. A. M.

1. On P.L. de Rham theory and rational homotopy type. Mem. Amer. Math. Soc. 8 (1976), No. 179.

Brown, E. H.

1. Twisted tensor products. I. Annals of Mathematics 69(1959), 223-296.

Cairns, S. S.

1. Triangulated manifolds and differentiable manifolds. Lectures in Topology. University of Michigan Press (1941), 143-158.

Cartan, H.

1. La transgression dans un groupe de Lie et dans un espace fibré principal. Colloque de Topologie (espaces fibrés) Bruxelles 1950. Masson et Cie. (1951), 15-27.

Cartan, H.

1. Cohomologie réelle d'un espace fibré principal différentiable. Sem. Cartan 1949/50, Exp. 19, 20.
2. Algèbres d'Eilenberg-MacLane et homotopie. Sem. Cartan 1954/55.

Cartan, H. & Serre, J. P.

1. Espaces fibrés et groupes d'homotopie. I. Constructions générales, II. Applications. C. R. 234(1952), 288-290, 393-395.

Chen, K. T.

1. Reduced bar constructions on de Rham complexes. Algebra, Topology and Category Theory (1976), 19-32, Acad. Press.
2. Connections, holonomy and path space homology. Differential geometry (Proc. Symp. Pure Math. V. 27, Ptl. 1973), 39-52.

Deligne, P. Griffiths, P. Morgan, J. & Sullivan, D.

1. The real homotopy theory of Kähler manifolds. Invent. Math. 29(1975), 245-254.

Friedlander, E. Griffiths, P.A. & Morgan, J.
1. Homotopy theory and differential forms. Mimeog. Notes (1972).

Griffiths, P.A.& Morgan, J.W.
1. Rational homotopy theory and differential forms. Progress in Math. 16, Birkhäuser, Boston (1981).

Grivel, P.P.
1. Suite spectrale et modèle minimal d'une fibration. Thèse, Univ. de Genève (1975).
2. Formes différentielles et suites spectrales. Ann. Inst. Fourier 29(1979), 17-37.

Haefliger, A.
1. Whitehead products and differential forms. Differential topology, foliations and Gelfand-Fuks cohomology. Lecture Notes in Math. V. 652(1978), 13-24.

Halperin, S.
1. Finiteness in the minimal models of Sullivan. Trans. Amer. Math. Soc. 230 (1977), 173-199.
2. Rational fibrations, minimal models and fiberings of homogeneous spaces. Trans. Amer. Math. Soc. 244(1978), 199-224.
3. Lectures on minimal models. Publ. U.E.R. de Mathématiques, Univ. de Lille I(1977, 1981).

Husemoller, D.
1. Fibre bundles. Springer-Verlag (1966).

Kahn, D.W.
1. The existence and applications of anticommutative cochain algebras. Ill. J. Math. 7(1963), 376-395.

Koszul, J.L.
1. Sur un type d'algèbres différentielles en rapport avec la transgression. Colloque de Topologie (espaces fibrés) Bruxelles 1950. Masson et Cie. (1951), 73-81.

Lehmann, D.
1. Théorie homotopique des formes différentielles d'après Sullivan. Astérisque 45, Soc. Math. France, Paris (1977).

Liu, N.P.
1. Mod 2 programming and planar imbedding. Acta Math. Appl. Sinica, 1(1978), 321-329.

MacLane, S.
1. Homology. Springer-Verlag (1963).

Meier, W.
1. Rational universal fibrations and flag manifolds. Math. Ann. 258(1982), 329-340.

Milgram, R.J.
1. The bar construction and abelian H-space, Ill. J.Math. 11(1967), 242-250.

Milnor, J. & Stasheff, J.D.
1. Characteristic classes. Princeton Univ. Press (1974).

Moore, J.C.
1. Algèbre homologique et des espaces classifiants. Sém. Cartan 1959/1960, Exp. 7.

2. Comparaison de la bar construction à la construction W et aux complexes K(π,n). Sém. Cartan 1954/1955, Exp. 13.

Quillen, D.

1. Rational homotopy theory. Annals of Math. 90(1969), 205-275.

Rashevskii, P. K.

1. Real cohomology of homogeneous spaces. Usp. Mat. Nauk 24:3 (1969), 23-90.

Smith, L.

1. Homological algebra and the Eilenberg-Moore spectral sequences. Trans. Amer. Math. Soc. 129(1967), 58-93.

Steenrod, N. E.

1. Milgram's classifying space of a topological group. Topology 7(1968), 349-368.

Sullivan, D.

1. Differential forms and the topology of manifolds. Manifolds Tokyo 1973, Univ. of Tokyo Press (1975), 37-50.
2. Cartan-deRham homotopy theory. Colloque analyse et topologie en l'honneur de H. Cartan. Astérisque No. 32-33(1976), 227-253.
3. Infinitesimal computations in topology. Publ. Math. 47(1977), 269-331.

Swan, R. G.

1. Thom's theory of differential forms on simplicial sets. Topology 14(1975), 271-273.

Thom, R.

1. Opérations en cohomologie réelle. Sém. Cartan 1954/1955, Esp. 17.

Thomas, J. C.

1. Fibrés minimaux algébriques et transgression (Part 1). Publ. UER de Mathématique, Univ. de Lille I(1976).

Vigné-Poirrier, M.

1. Quelques problèmes d'homotopie rationnelles. Thése, Univ. de Lille I(1978)

Weil, A.

1. Sur les théorèmes de deRham. Comm. Math. Helv. 26(1952), 119-195.

Whitehead, J. H. C.

1. On C^1-triangulations. Annals of Math. 41(1940), 809-824.

Whitney, H.

1. Geometric integration theory. Princeton Univ. Press (1957).

Wolf, J.

1. The cohomology of homogeneous spaces. Amer. J. Math. 99(1977), 312-340.
2. The real and rational cohomology of differential fibre bundles. Trans. Amer. Math. Soc. 295(1978), 211-220.

Wu, Wen-tsün

1. A theory of imbedding, immersion and isotopy of polytopes in an euclidean space. Science Press, Beijing (1965); Chinese version (1978).

2. Layout problems in printed circuits and integrated circuits. Appendix in Chinese version of [1], 213-261.

3. A new functor in algebraic topology. (Chinese), Kexue Tongbao 9(1975), 311-312.

4. Theory of I^*-functor - The real topology of homogeneous spaces. (Chinese), Acta Mathematica Sinica 18(1975), 162-172.

5. Theory of I^*-functor in algebraic topology - Real topology of fiber squares. Scientia Sinica 18(1975), 464-482.

6. Theory of I^*-functor in algebraic topology - Effective calculation and axiomatization of I^*-functor on complexes. Scientia Sinica 19(1976), 647-664.

7. (with Wang, Qui-ming) Theory of I^*-functor in algebraic topology - I^*-functor of a fiber space. Scientia Sinica 21(1978), 1-18.

8. On calculability of I^*-measure with respect to complex-union and other related constructions. Kexue Tongbao 25(1980), 185-188.

9. de Rham - Sullivan measure of spaces and its calculability. The Chern Symposium 1979, Springer-Verlag (1980), 229-295.

10. A constructive theory of algebraic topology - Part I. Notions of measure and calculability. J. Sys. Sci. & Math. Sci. 1(1981), 53-68.

11. The de Rham theorem from a constructive point of view. Proc. 1981 Symp. Diff. Geometry and Diff. Equations (Shanghai-Hefei), Science Press (1984), 497-528.

I N D E X

MIX
Papier aus verantwortungsvollen Quellen
Paper from responsible sources
FSC® C105338

If you have any concerns about our products,
you can contact us on
ProductSafety@springernature.com

In case Publisher is established outside the EU,
the EU authorized representative is:
Springer Nature Customer Service Center GmbH
Europaplatz 3, 69115 Heidelberg, Germany

Printed by Libri Plureos GmbH
in Hamburg, Germany